Praise for

Team Guide to Metrics for Business Decisions

"I love this book. It gets straight to the point providing easy to understand, and practical guidance on how to use data to answer all the commonly asked questions. Questions such as 'How do we know this works?' and 'When will it be done?'. Highly recommended to anyone who works with a team looking to improve their ability to build and operate software systems."

— Amy Phillips, Engineering Manager at Gousto

"A beautifully written book about metrics that answers real key questions that teams face daily. With clear explanations, examples, myths and getting started tips this should be anyone's go to book for making metrics easy and informative."

— Helen Meek, Coach, Trainer & Consultant

"Despite my 20 years of Agile experience the book surprised me with a brand new and fresh perspective on the metrics matter! Each topic is thoroughly covered and explained with great clarity, nothing is taken for granted, and many false myths will be dismantled.

You will understand deeply the essence of 'Why' and 'How' a metric should be adopted and 'Where' to focus on to get real benefits like eliminating bottlenecks, making good decisions and providing realistic forecasts. This book is a 'must' for your Agile library!"

— Stefano Luzi Crivellini, Chief Delivery Officer at Creactives S.p.A.

"Software development teams need to go beyond estimation in order to forecast when work will be done. They also need to prove that their software is valuable for customers. Mattia and Chris's book is an excellent guide for how to have data-driven conversations, and it provides many tools and resources for success."

— Heidi Helfand, Author of Dynamic Reteaming. *Director of Engineering, Excellence at Procore Technologies*

"Mattia and Chris guide us through the questions that we've all been asked: How fast are we going? How long will this take?

Full of real examples and suggestions they share their experience in a direct and actionable way. A book to read and to have on our desk to take decisions based on better conversations generated by data."

— Stefania Marinelli, Agile Manager at Hotels.com

"I love that this is written specifically for team members, rather than tech leadership - there are questions that all teams need to ask themselves about the flow, quality and value of their work. The clear examples and explanations in here will spark curiosity in any team member and equip them to use their own data to help them focus on what matters most."

— Victoria Morgan-Smith, Director of Delivery, Internal Products at Financial Times

"Team Guide to Metrics for Business Decisions feels like a long overdue publication that has been missing from the Lean-Agile world. Whilst there can be no doubt about the wealth of useful material on metrics that is out there for practitioners to consume, I've always felt like there was something missing. ... Through reading this book it's safe to say my opinion has changed.

I would consider this book essential reading for people new and/or experienced in the lean-agile world. The questions it answers, the perspectives it gives and stories told make it a book with great learnings and practical guidance for everyday application in your organization."

— Nick Brown, Agile Lead (IFS) at PwC UK

Team Guide to Metrics for Business Decisions

Mattia Battiston and
Chris Young

Conflux Books
Leeds, UK

Team Guide to Metrics for Business Decisions

Mattia Battiston and Chris Young

Published by Conflux Books, a trading name of Conflux Digital Ltd, Leeds, UK.

Commissioning Editor: Matthew Skelton
Cover Design: Elementary Digital and Matthew Oglesby

For information about bulk discounts or booking the authors for an event, please email info@confluxbooks.com

ISBN	978-1-912058-81-5
eBook ISBN	978-1-912058-63-1
Kindle ISBN	978-1-912058-80-8
PDF ISBN	978-1-912058-65-5

Conflux Books: confluxbooks.com

Contents

Team Guides for Software

Pocket-sized insights for software teams

The *Team Guides for Software* series takes a *team-first approach* to software systems with the aim of **empowering whole teams** to build and operate software systems more effectively. The books are written and curated by experienced software practitioners and **emphasize the need for collaboration and learning, with the team at the center**.

Titles in the *Team Guides for Software* series include:

1. *Software Operability* by Matthew Skelton, Alex Moore, and Rob Thatcher
2. *Metrics for Business Decisions* by Mattia Battiston and Chris Young
3. *Software Testability* by Ash Winter and Rob Meaney
4. *Software Releasability* by Manuel Pais and Chris O'Dell

 Find out more about the *Team Guides for Software* series by visiting: http://teamguidesforsoftware.com/

Acknowledgements

We've learnt and taken huge inspiration from many experts and practitioners whilst writing this book. We would like to thank the following people for their help and inspiration (in alphabetical order): Alexei Zheglov, Andy Carmichael, Bazil Arden, Dan 'KanbanDan' Brown, Dan Vacanti, David J. Anderson, David Lowe, David Shrimpton, Dimitar Bakardzhiev, Donald Reinertsen, Emily Webber, Gaetano Mazzanti, Håkan Forss, João Miranda, Johanna Rothman, John Cutler, Karl Scotland, Larry Maccherone, Melissa Perri, Mike Burrows, Pawel Brodzinski, Prateek Singh, Robin Weston, Sam L. Savage, Seb Rose, Simon Machin, Steve Smith, Troy Magennis, Vasco Duarte, Woody Zuill.

We also give a big "thank you" to our reviewers: Amy Phillips, Clare Sudbery, Don Maclellan, Heidi Helfand, Helen Meek, Nick Brown, Stefania Marinelli, Stefano Luzi Crivellini, Victoria Morgan-Smith, Peter Marriott.

We owe a great deal of gratitude to our publisher, Matthew Skelton, and our editor, Manuel Pais, for their guidance and patience.

Mattia would also like to thank his family - Aindrea, Anthony and William - for their support and patience over the many days and nights that he spent working on this book, and the people in the NIM team at Sky, who over the years have inspired and supported many of his experiments.

Chris too would like to thank his family - Toni-Marie, Elco and Louie - for their support, Kate Gray for being a great friend and mentor and James Carpenter, Craig Russill-Roy and Andrew Fox from Honeycomb.

– Mattia Battiston and Chris Young

Index of Case Studies

Introduction

"Why should the team care about metrics?"

How can software teams keep a laser-like focus on outcomes whilst improving practices and flow? How can software teams understand where the bottlenecks are in their processes? And how can software teams provide reliable forecasts of when work will be done? The answer to these questions is: software teams should use Business Metrics. Teams that use Business Metrics understand Throughput, Lead Time, Forecasting, Flow, Quality, and Value - all measures that speak directly to business outcomes. By using Business Metrics, your software team will produce software that is more business-relevant with more certainty and less waste.

What is covered in this book

Our goal in this book is to equip teams with the means to help them **measure the value and quality of their work together with the time and resources required to get the work done.**

By doing this we hope to make it easier for teams to work on the things that matter the most, both to the team and the wider business or organization for which they work, and do so at the right time.

To be able to measure you need metrics. Metrics give teams a shared, tractable, resource which can be poked and prodded, questioned and explored. From these metrics come conversations and insights that we have found help teams succeed.

What we are not trying to do is tell teams how to do the work. This is not a methodology book. It's not about how to use Kanban, Scrum or whatever your preferred way of working is. Rather it is about how to instrument the work and its outcomes in order to make both better for everyone involved.

All the time in the world

> Time is an executive's scarcest and most precious re-
> source. And organizations - whether government agen-
> cies, businesses, or nonprofits - are inherently time
> wasters.
>
> – Peter F. Drucker The Effective Executive

As we and our teams build and run software we are continually asking and being asked questions:

- How much do we have left to do this week/month/year?
- How much time do we have to finish this?
- How long does it take us to deliver things?
- How much does it cost per month to operate our software?
- What should we do next?

These questions all have one thing in common. They are about our use of and the value of time. To help us answer these questions we need to measure. And we must always measure over time.

For example, say we are running an e-commerce website and we measure how many people are using our website at any given time. Just knowing that there are 1000 people using our website right now isn't particularly useful.

If, on the other hand, we know that we usually have 1000 people at a time between nine in the morning and five in the evening but that this number drops off to only 100 overnight we can use that information to decide how much computing resources we want to put in place at what time of the day.

Or, to take another example, we notice that over the past few weeks the number of defects being logged against our system has been going up at a steady rate, from two defects/week to four, then six and now eight defects/week. Then we know we have a trend that we should be worried about.

This is an actionable book primarily for teams as time is the *entire* team's scarcest and most precious resource. Time is wasted if we don't know what we should be working on and when.

 We can quantify the cost of delay by measuring the **operational effectiveness** of our work, its **quality** and the **value** it creates.

Value, quality, & operational metrics

We can split metrics into three tiers based on their purpose. Each tier supports the ones above it:

Metrics pyramid: value, quality, operational

At the bottom we have operational metrics. These are the underlying 'hygiene factors' that indicate the ways of working required to produce the value and quality outcomes required: throughput, flow, variability, and so on. These metrics cast a light into otherwise dark corners and give us situational awareness.

Next comes quality. We want to be able to measure the quality both of the work we produce and also how we produce it (how we operate). The quality of the work we do affects both the operational effectiveness and its value. If our code is full of bugs then this prevents people using it from getting value out of it. Plus we spend more of our time fixing those bugs rather than delivering valuable new features.

At the top is value. This is where we want to end up, being able to measure the value of our work - the outcomes and results it produces. Otherwise, how do we know it is worth doing and spending time on in the first place?

Operational metrics

We start at the base of the pyramid with operational metrics.

We need these to be able to answer questions like:

- How much do we have left to do this week/month/year?
- How much time do we have to finish this?
- How long does it take us to deliver things?

To answer these questions we need metrics derived from how the team works, as opposed to their output. We measure things like:

- Delivery rate - also called throughput. How fast is the team delivering the work?
- Cycle time - also called lead time. How long does it take the team to complete the work?

We suggest that teams start by looking at their operational metrics as these are within the team's *Zone of Control*. They can then be used as an example to those within the team's *Sphere of Influence* (Adzic2014) such as customers, product managers or heads of business.

Quality metrics

Once we can see our operational metrics, we are in a position to look at the quality of the work we are doing. Like operational concerns, quality is within the team's zone of control. As David Anderson put it "[quality] is a technical discipline that can be directed by the function manager" (Anderson2010). It is something that as a technical or engineering team *you* own.

Raising the quality of your work - "getting your own house in order" - puts you in a good position to have fruitful conversations with your customers and collaborators, and is a way of strengthening trust with them.

We need to be able to answer questions like:

- Does our software consistently do what people expect it to do?
- Is our software available at the times and in the places where people need it?
- When things go wrong with our software, what does it take to get them fixed?

A way of qualifying, quantifying and then measuring quality is to think of it in terms of what John Seddon calls "failure demand":

> Value demand is 'demand we want', demand that the service is there to provide. Failure demand is demand caused by a failure to do something or do something right for the customer.

– John Seddon "Systems Thinking in the Public Sector"

For a team building and running software, failure demand is caused either by the defects in the software put into production or by building something that doesn't fulfil the customer needs, that is something that does not provide value.

Value metrics

In our experience there is rarely a shortage in the supply of ideas for things to do with a team's scarce time. We need to know which of those are going to create value.

This means we need to be able to answer questions like:

- What should we do?
- When should we do it?
- What should we do next?

To answer this we need some measure of the worth of what we seek to achieve and the degree to which our actions are having the desired outcome. These are our *value metrics*.

For example:

- How many customers have we gained this month?
- How much are they spending?
- Which features are they using the most?
- How long do people stay with us as customers?
- Do they recommend us to their friends?

Value metrics help to provide a common goal for the team by emphasizing the purpose and outcomes of the software, reducing the time needed for decision making and avoiding time wasted on interesting but less valuable work.

How to use this book

The metrics we generate are not an end in themselves. They are there to answer questions and to enable the team to make better business decisions. Each chapter of this book poses one of the questions we hear the most when working with teams:

- Chapter on Throughput - How fast are we going?

In the fast-paced world of software development we have found that it is all too easy to lose track of how much you can actually get done in say a week or a month. When demand from your customers is piling up, by looking at the operational metric 'Throughput' you can get a handle on how well balanced your capability is to deliver against this demand.

- Chapter on Lead Time - How long will this take?

How long is a piece of string? Software is complex and requirements are emergent. Both customers and the team want to know how long a piece of work is going to take to complete. By looking at the operational metrics "Lead Time" and "Cycle Time" we show you how to answer this perennial question.

- Chapter on Forecasting and planning - When will it be done?

Now we're motoring. We have a predictable delivery rate and can give our customers realistic expectations of when individual pieces of work will get done. When it comes to bigger chunks of work though, like epics or projects, questions remain: When will it be done? How much can we do by a particular date? In this chapter we look at how to make forecasts that help us take those decisions with confidence, dealing with uncertainty and highlighting risks early on.

- Chapter on Flow - Where's the bottleneck?

As we look at queues of work alongside work that is in progress we will see that some of the queues are longer than others. Where these 'bottlenecks' appear tells us a lot about how we are working and what we can do to improve it.

- Chapter on Metrics for Quality - How do we know this works?

There is no point knowing how quickly we can deliver 'stuff' if that stuff doesn't work. In this chapter we look at the concept of "Failure Demand". We see how we can measure it and look at ways to remove it from the equation.

- Chapter on Metrics for Value - Why are we making this software?

Value metrics help us to answer the question "Why are we doing this?". You could even decide to 'Start with Why' as Simon Sinek famously said (Sinek2011) and read this chapter first. The other chapters are all about "getting the house in order" - doing a great job. Value metrics help us to take this practice of instrumenting and reflecting upon our work and apply it to the outcomes and purpose of the work itself. We can start measuring the impact of our work: by choosing to measure impact we prompt conversations about purpose ("why?").

Get Started!

Each chapter is readable independently, containing the necessary level of detail to be understood and actionable on its own, without requiring any of the other chapters in the book to be read first (although certainly reading the full book will provide a more comprehensive understanding of the concepts and practices and their inter-relations).

We've structured the book in an order that is intended to take you and your team from a practice with no metrics to one where metrics are driving your day-to-day decisions about the work.

That doesn't mean you have to read it linearly cover to cover. This is a book for teams and we suggest you start with a conversation with the other members of your team about what is and what is not working for you.

Ask them the questions above - 'How fast are we going?', 'What are we doing next?', 'Why are we making this software' etc. - and see how much interest you get and what answers people have. Which question do they most want answering? Our thesis in this book is that these questions are best answered by using metrics. By working through the chapters with your team you should be able to discover more about the work you are doing and better answer the questions.

Feedback and suggestions

We'd welcome feedback and suggestions for changes.

Please contact us at:

- email: info@bizmetricsbook.com
- Twitter: @BizMetricsBook
- Leanpub: leanpub.com/metricsforbusinessdecisions/feedback

Mattia Battiston & Chris Young

1. Throughput: How fast are we going?

Key points

- **Most teams have a need for predictability,** they need to answer questions like "How much work can we complete? How long will it take?"
- **Story points and velocity often give us little predictability**
- **Use throughput** (number of work items completed in an iteration) **to know how fast the team is going**
- **Use throughput to improve planning and forecast** how much work can be completed in the next iteration
- **Check the throughput trend to know if the team is improving**

This chapter focuses on **throughput**, the number of work items that were completed during a time period.

1.1 Why should I care about throughput?

As a **team member**: throughput helps you become more predictable, so you are able to make more realistic commitments. You have data to fight the pressure to fit more work into the available time and thus avoid death-march projects (DeathMarch).

Using throughput you also spend less time estimating, which leaves more time for other work. Meetings should get faster as decisions become less based on personal opinions and more on data.

As a **Scrum master** or **coach**: throughput helps you check if the team is improving and validating the effectiveness of experiments. Facilitating planning meetings and estimation sessions becomes much easier as you focus on the content of the stories rather their size.

As a **product owner**: throughput gives you realistic expectations of what the team can and can't achieve. You start thinking in terms of "What are the chances that we can fit this story in the next couple of weeks?" rather than relying on the team promising that "they'll try".

As a **manager**: throughput helps teams make realistic promises so you'll know when they're likely to start/finish work. You'll have data to play what-if scenarios.

As a **customer**: throughput gives the team the ability to make you realistic promises. Disappointments become the exception rather than the norm. You can be confident that when the team says they'll get something done, they probably will.

1.2 What's the problem with velocity and story points?

This chapter is all about answering questions like "How fast are we going?" or "How much work can we complete?". Typically Agile teams use story points and velocity to come up with an answer.

Story points are a measure of the perceived size of a user story. For example 'Copyright on webpage footer' might be considered small and thus allocated a single story point, whereas 'Enforce password strength policy' might be considered large and so be given ten story points.

Teams then apply their *velocity* - that is the number of story points per sprint the team can deliver - to decide how many stories they can include in the next sprint. They might also use velocity to determine how many sprints it will take to complete a given set of user stories (Cohn2016 Cohn2014).

These practices are so common amongst Agile teams that they are considered the norm. However, it's not all roses: let's look at some of the problems with this approach.

The problem: low correlation with lead time

When Mattia first joined his team in late 2013, they were playing planning poker (PlanningPoker) and estimating using story points. The team was used to it and had been doing that for a long time. Mattia wondered if their estimates were making sense and if they were at all predictable, so the team started to gather some data. Little did they know that they were in for a big surprise...

The chart below shows the results found by Mattia and his team. In particular, for each story, we can see the relationship between estimated story points and how long it actually took to complete the story (in other words, the story's lead time).

Mattia and his team were absolutely shocked to discover that 1-point stories were taking between two and ten days, 2-point stories were taking between two and twenty days, and 3-point stories were taking between six and eighteen days! There was very low correlation between their estimated story points and the actual lead time.

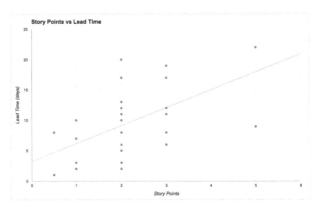

Very low correlation between story points and actual lead time

Such low correlation meant the estimates were adding very little value: the team really had no meaningful way of predicting how long a story would take based on its estimated points. Consequently, their velocity made little sense and the team was highly unpredictable.

The team therefore decided that it wasn't worth it to continue investing time estimating using story points anymore. Instead, they started to base predictions on historical data and the metrics described in this book.

In particular, throughput proved to be the metric that could answer their questions in a way that was easier, cheaper and more accurate.

What causes these problems?

There are three important factors that have a much higher impact on lead time than user story size, and that when left unmanaged make our teams throughput unpredictable.

First, do we have a **high amount of work in progress (WIP)**? When we work on too many things at the same time we are not able to focus on finishing the tasks that are already in progress. We waste time in context switching, the quality of our work decreases, and even stories that appear to be simple end up taking longer than expected.

Second, are the **queues in our process** visible? How long does work sit in those queues? Very often in our processes there is some waiting time between one activity and another. For example: waiting for a developer to be free to start a story, waiting for someone to be free to test a story, waiting for the next release, etc. These queues are often invisible, they're not represented on our boards, and it's really common to ignore them when we estimate. When these queues are not managed they lead to a lot of work in progress but on hold, which in turns lead to high unpredictability.

Third, are **unexpected events** common? Events like being blocked by a broken test environment, having to clarify requirements, or 'urgent' problems such as a defect in production, make us interrupt and put on hold activities that have already been started.

When we have one or more of these problems - and most teams we've worked with certainly seem to - then the estimated story size doesn't matter as much as it has a very low correlation with the actual time taken to complete the work.

Even the simplest story is going to take ages if we're not focused, if we waste time in numerous hand-offs, or if it's constantly overtaken by more urgent activities. For more details on solving these problems see the chapter on Forecasting and planning.

Are story points bad?

The use of story points, or rather the non-use of them, has been a controversial topic in the Agile community. So let us be clear: by no means are we trying to discredit story points in this book. For many teams and companies they have represented a tremendous step forward from the past, and they're still using them successfully.

What we **are** saying is that if you recognize some of the problems described above, then chances are that story points are not giving you much predictability.

Several people in the community have found similar results to ours:

- Vasco Duarte, one of the main proponents of #NoEstimates, found high correlation between story points and number of completed stories, suggesting that they're measuring the same thing (Duarte2012).
- Larry Maccherone analyzed data from 1000s of teams and found that throughput was the best measure for productivity and predictability (Maccherone2014).
- Ian Carroll found low correlation between story points and lead time for 25 teams (Carroll2016).
- Folks at ThoughtWorks found that they were getting the same predictability using story count (throughput) or story points (ThoughtWorks2013).
- Nader Talai presented very similar results, where throughput gave his team the same predictability as story points (Talai2014).
- Pawel Brodzinski, a bit provocatively, created a new set of planning poker cards where the only values are "1", "Too Big", "No Clue". For him the only interesting question is whether the story is small enough, without needing to estimate whether it's a 5 or an 8 points story (Brodzinski2015a).

Should I stop using story points and velocity?

By all means, if you're finding story points and velocity useful in your team please keep using them! What we invite you to do however is to ask yourself and your team: "Are our estimates really working?".

Whatever estimation process you use at the moment - story points, t-shirt size (Singh2016), ideal days (Rao2014), or others - generate a chart like the one above. Is there a correlation between your estimates and the time taken to complete your stories? You might be surprised by the results.

Our advice is: don't drop whatever estimation process you're currently following, but start collecting data and generate some metrics. After a while you can compare your estimates and your metrics and decide what makes you more predictable. You should be able to start using your data to improve your estimates.

And if at some point you end up deciding that you want to drop story points, then having team data in your hands will make it a lot easier to convince the people around you.

1.3 What is throughput?

Throughput (often also called delivery rate, or story count) is the number of work items that were completed during a particular period of time.

Throughput is extremely simple to calculate and provides high value, that is why we think it makes for a great starting point.

How do I calculate throughput?

It's extremely simple: just count the number of work items, such as user stories, that have been completed in a fixed period of time. The period could be a two-week sprint, a week, a month, and so on. The important thing is to keep it consistent. If you do Scrum, or work with iterations, the easiest thing is to make it coincide with the length of your sprint. Otherwise, just pick a cadence that works for you.

For example, say our team chooses a two-week period. Then every two weeks we count how many stories have been completed in that period of time.

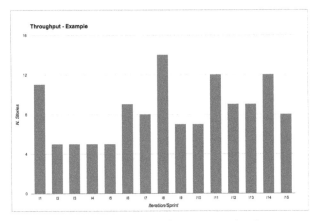

Example throughput - Number of stories completed per sprint

Tips for calculating throughput

If you use a *physical board*: you can simply count the number of cards that have reached your "Done" column (or whatever you consider as completed, according to the policies of your process). After you've counted the stories you might want to clean your "Done" column, so that next time it will be easier to see what has moved there since.

If you use an *electronic board*: some tools let you extract this information. If yours doesn't, you can quite easily calculate the throughput just by knowing the end date of each work item. Simply calculate what iteration/sprint that end date falls into. For an example, have a look at our public repository (https://github.com/ SkeltonThatcher/bizmetrics-book) or at Troy Magennis' collection of spreadsheets (MagennisSpreadsheets).

Split by work item type

We recommend splitting throughput by *work item type* in order to know how many of each type we usually get done in a given period.

This will enable us to make much better predictions (more on this in chapter 6).

Different work item types are usually characterized by:

- Whether they follow a *different process* (for example, some user stories might not need to be tested, whilst some have to go through a special test environment)
- Follow the same process but at a consistently *different speed* (for example, work we do in our legacy codebase is consistently much slower than work on the greenfield app).

Don't go overboard though: typically we'd expect teams to only have a handful of work item types.

In Mattia's team they use:

- "Dev Stories" to represent "normal" work. These stories go through each step of the process before being "done" - they are analyzed, developed, tested, and then released.
- "Direct Stories" to represent work that goes straight from "development" to "done", like writing a report, investigating a support issue, responding to a team email. These stories follow a different process as they skip the test and release steps, and therefore have a much lower lead time than "Dev Stories".
- "Release Testing Automation" to represent work that the testers are doing to automate their manual regression tests. These stories follow the same process as "Dev Stories" but are usually much quicker.

Most metrics only make sense when analyzed for a particular work item type:

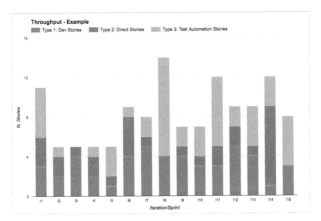

It's useful to split the throughput by work item types to make more accurate predictions

For example, say that we are working in iterations and we're dedicating one sprint to stories on a new technology that we're not familiar with. We don't have historic data for these, so we're unsure how long it will take. Our gut feeling is that it will take longer than usual, thus the throughput in this sprint will be quite different from other sprints.

It's important to have the data to acknowledge that the throughput change was due to this new type of work. We can then decide if we should treat this sprint as an outlier or if we need to revisit our forecasts instead.

Case study: differentiating work items

João Miranda, Engineering Manager at a large European bank, writes:

We went agile 7 years ago, starting with a small number of teams going all-in for Scrum as our base framework. These teams were formed to execute strategic projects, so they had full attention from their (internal) clients and did not have to deal with production issues (maintenance teams would handle those).

Once we started scaling out Scrum to all the development teams (about 70), we faced new challenges. Today development teams have multiple responsibilities: new features; maintenance and bug fixing; refactorings and optimizations; urgencies (unplanned work). Balancing all these needs while keeping the sprint scope stable is a hard problem.

The approach we took, which serves us well to this day, is simple. We split each team's available capacity into bands. A band is a percentage of the available effort in a sprint. The "new features" band includes all scenarios that provide new capabilities to users, regardless of feature size. Enabling a new loan product, even if it's just some parameterization in a database? New feature. The "maintenance and bug fixing" band includes all fixes in production and configuration changes (for example changing loan rates). The "refactorings and optimizations" band includes all the work the team does to explicitly reduce the size of the "maintenance and bug fixing" band (in short, paying down technical debt).

Finally, "urgencies" is an explicit buffer to handle all unexpected activities that happen during a sprint. In this way, we help the team to avoid overcommitment and keep the sprint plan mostly stable.

The allotted size of each band is unique per team and may change if the team's dynamics changes. For instance, if a team has a lot of maintenance work, chances are its "maintenance and bug fixing" band is larger than average. But its "refactoring and optimizations" band will also be larger so that the team can bring back things under control.

We don't have any default or baseline for the size of each band, context is king.

1.4 Use throughput for short term predictions

"How many stories can we complete in the next sprint?"

"How much can we get done in the next two weeks?"

These are common questions most teams get asked during sprint planning (in Scrum), queue replenishment (in Kanban), or simply when talking to stakeholders.

The good news is that the team can answer them by looking at their throughput stats.

THROUGHPUT STATS

	Total	Type 1: Dev Stories	Type 2: Direct Stories	Type 3: Test Automation Stories
Average	8.4	3.2	1.9	3.3
Median	8.0	3.0	1.0	3.0
Std Deviation	2.9	1.3	2.0	2.6
Min	5.0	1.0	0.0	0.0
Max	14.0	5.0	8.0	10.0
Mode	5.0	4.0	1.0	1.0
5% percentile	12.6	5.0	5.2	7.9
25% percentile	10.0	4.0	2.0	4.5
50% percentile	8.0	3.0	1.0	3.0
80% percentile	5.0	2.0	1.0	1.0
95% percentile	5.0	1.0	0.0	0.7

These stats about our throughput help us make short-term predictions

These stats are telling us a lot of useful information:

- Average (mean) and median: how many stories do we complete during a sprint, typically?

  (by the way, we usually prefer to look at the median rather than the mean average as it's less impacted by outliers, so it's usually a more accurate representation of reality)

- Standard deviation: how much variability in throughput have we got?

  (we usually look at its trend to see if variability is increasing/decreasing. The actual value is not important as it's only reliable when our data follows a Gaussian distribution, which is rarely the case in a software development process (Zheglov2014a))
- Min and max: number of stories ever completed.
- Mode: what's the most common throughput?

  (this is the number that occurs most often, so it's the one that people usually remember off the top of their head)
- Probability percentiles: what's the likelihood of completing that many stories in one sprint?

Let's focus on stories of type 1 for a second ("Dev stories" in our example) and look at their percentiles. The 80th percentile tells us that in 80% of sprints we complete at least two "Dev Stories". Similarly, looking at the 50th and 25th percentiles we know that in 50% of sprints we complete 3 stories, and only in 25% of sprints we complete 4 stories. Therefore if we want to have a high level of confidence that the "Dev Stories" we start working on will be completed within two weeks, we should only pull in the next 2-3 "Dev stories".

We could risk it and pull in a fourth one, but the chances of getting it done are very slim - 25% chance, to be precise - so we wouldn't want to make any promises to our stakeholders about that fourth story!

If we now look at type 2 ("Direct Stories"), we know that in 80% of sprints we also complete one "Direct Story" on top of the other "Dev Stories". So we can also safely pull in one of these stories. Similarly, we can confidently expect to complete one or two "Test Automation Stories".

In short, for each story type, we look at the throughput of that type of work in order to make predictions about how many stories we can complete in the next period.

Notice how this approach makes planning sessions a lot faster: since we know that we rarely complete more than four stories, there's no point talking about stories that we won't even look at.

Queue replenishment: "How often should we decide what to work on?"

In our experience working with Scrum teams, a common struggle we see is planning for the next two weeks.

Regardless of the decisions taken during sprint planning, the team's plans often get trumped by new urgent stories, emergencies, and stakeholders changing their mind.

The team feels that they're not reactive to change, and that they should be more flexible changing scope or priorities. That's a symptom that their batch size of two weeks is too large for them, and they should instead plan shorter periods of time. And so they find themselves asking questions like "How often should we do planning? How many stories should we have in our 'Next' column?".

Don Reinertsen (Reinertsen2009) teaches us that reducing batch size is almost always a good idea, but where is the sweet spot? The more often we plan the more flexible we can be, but we also don't want to spend all our time planning, and the people we need to make those decisions might not always be available. Luckily, we can use our throughput to decide how often we should replenish our queue.

Let's have an example: say that our team would like to plan every week instead of every two weeks, meaning that every week we have a chance of reviewing priorities and changing direction. How many stories should we put in our 'Next' column? Well, let's use our throughput stats: in a week we usually complete three to four stories. So let's have four stories in our 'Next' column, and every week we will replenish it.

At Honeycomb TV the throughput of the technology team, which uses Kanban, is such that a queue of four items empties out once a week. Hence every Tuesday a Google Hangout takes place where the technology team and its customers, the wider business, gather around the Trello board and decide what to put into the queue next.

Queue Replenishment

In the screenshot above there are two pieces of value work - that is features (green label) - and two pieces of operational overhead (black label).

When deciding how often to replenish the queue, you need to take into consideration the product owner's availability. If the product owner is needed for planning but is only available every three days that means that we can't possibly replenish more frequently than every three days. We'll size our queue based on how many stories we usually complete in three days.

Queue replenishment: "Are we reducing technical debt?"

We can use throughput to analyze the percentage of effort we spend on work items of each type. A typical usage would be to make sure that we are working on technical debt items: we can agree as a team on a percentage of capacity to reserve for technical debt, and honor it at queue replenishment.

For example, Chris's team realized that they were spending too much effort fixing defects, taking away time from other value-adding activities. Recently they had been focusing too much on adding new features and weren't doing enough to keep technical debt under control, resulting in a high number of issues.

They decided on a a new team policy: whenever they were pulling in new work, for every three business stories they would pull in a technical story.

Using this policy made sure that the team wasn't pressured into working only on business work and that technical debt was being addressed.

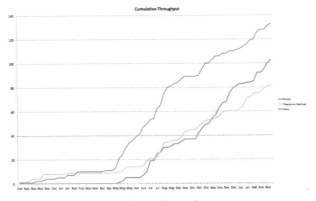

Cumulative Throughput

Visualizing predictions with rolling wave forecasting

A nice way to visualize our predictions is by using a technique called "rolling wave forecasting", introduced by Vasco Duarte in his book "No Estimates" (Duarte2015).

We visualize the stories in our backlog using different colors. Each story gets a different color based on how confident we are that they will be completed in the next two weeks:

- green = definitely
- yellow = probably
- orange = possibly
- red = unlikely

ROLLING WAVE FORECASTING

Backlog (prioritised)	Will complete in next 2 weeks?
Story 1	Definitely
Story 2	Definitely
Story 3	Definitely
Story 4	Probably
Story 5	Probably
Story 6	Probably
Story 7	Possibly
Story 8	Possibly
Story 9	Unlikely
Story 10	
...	

Legend	
Definitely	> 80% confidence
Probably	50% - 80% confidence
Possibly	25% - 50% confidence
Unlikely	< 25% confidence

Rolling wave forecasting for visualizing our predictions

These levels of confidence are taken from our data, using the percentiles from our throughput. If the 80th percentile is three stories, then we're 80% confident that the first three stories in the backlog will be completed in the next two weeks (green). We're 50% confident that we'll complete up to six stories, so stories four to six are yellow. And so on for orange and red.

1.5 Use throughput to validate experiments

"Are we improving?"

"Are we going faster/slower?"

One of the Kanban practices says "evolve experimentally" (AndersonCarmichael2016). Metrics are a fantastic tool to help us do that: we drive our continuous improvement through running experiments, and metrics help us decide whether those experiments are successful or not. Whenever we introduce a change, for example implementing an action from a retrospective, we rely on our metrics to know if the change is helping us improve.

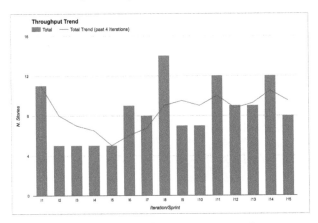

Throughput trend - are we improving?

Throughput is great to know if we're going faster or slower. In particular, we look at the throughput trend by calculating the throughput average over the past four iterations.

We look at the throughput trend rather than absolute values because continuous improvement is about going in the right direction, rather than hitting a particular target. So as long as the trend is going up, we're improving.

What would an experiment look like?

The key part of an experiment is treating our ideas as hypotheses and setting our expectations before we start: what do we expect to happen after we introduce this change? How and when are we going to know if this experiment has been successful or not?

For example, say that during a retrospective we come up with the idea that we should do more pairing between developers and testers because we think it will reduce the amount of rework due to bugs. Rather than just saying *"let's do more pairing"* we can formulate it as an experiment that might look like this:

Current situation: Several stories need rework due to bugs found during exploratory testing. Average of 0.5 bugs per story.

Our hypothesis: We believe that if developers and testers paired more, developers would learn how to prevent those bugs, resulting in less rework, and therefore having more time to complete other valuable work.

Action: For the next four sprints, each developer will spend at least one day a week pairing with a tester.

Expected results: Initial slow down in the first sprint due to time invested in learning about testing strategies, then in the last sprints we expect less bugs and less rework, therefore more stories completed.

How to measure our results: In the first sprint we expect throughput to go down of 1 or 2 stories. But in the following sprints we expect less bugs and higher throughput (extra 1-2 stories) compared to now.

Notice that we're not just checking that the number of bugs goes down, as that could be achieved by simply spending more time on a story. As well as less bugs, we want the throughput to increase to make sure that we're actually improving.

This is just an example structure for your experiments. There are many techniques that can help here, like A3 thinking (A3Thinking) or the Toyota Kata (Rother2009 Forss2016).

Your experiments don't have to be as formal as this, but make sure to talk about what you expect before changing something, and then use your metrics to determine success, in particular analyze your throughput changes.

How do I know if my experiment passed/failed for the right reasons?

In our experiments we do our best to change only one variable at a time, to try and find a cause-effect relationship between our change and the results. Nevertheless, it's impossible to fully isolate an experiment. Real life is complex (as in Cynefin-complex (Cynefin)) and every day there are many things that influence the outcome of our work. So how do we know if an experiment was really successful, and that achieving the expected results was not just a matter of chance?

The best advice we can give is: use multiple metrics. Never rely on a single metric for checking your results, as a single metric is easily influenced by many other factors than your experiment, including people trying to cheat or 'game the system'.

For example, a common mistake is to set a particular velocity as a target. This very often results in the team estimating each story with a few extra points - what used to be a five is now an eight, threes are now fives, and so on. By focusing on one metric (velocity) the team is unintentionally encouraged to optimize their behavior to reach that target, at the cost of cheating.

Instead, if you look at multiple metrics and observe that most of them have progressed, then it's much more likely that the experiment was the cause. Only when multiple related metrics are progressing are we really improving.

How much variation is normal?

Our teams are made of human beings working in complex environments, so some variation is perfectly normal. Actually, we'd be worried by a complete absence of variation: we'd interpret that as inflexibility and a lack of innovation.

But just how much variation is normal? Only you and your team can know. You're the only ones that know your context and can decide if something is a one-off outlier, rather than the symptom of a problem.

For instance, the "Throughput Trend" figure above shows that from iteration I6 onwards the throughput is oscillating around 8 stories per sprint. However, throughput for iteration I8 was considerably higher (14 stories). The team knew this was an outlier because they had received extra help from other teams. Variation is fine as long as you know what caused it.

Having said that, beware of tampering (Burrows2014): if you react to events too quickly you might be reacting to what is just normal variability, and end up making your system unstable.

Real-life examples

Increase WIP limit to prove that it's a bad idea

We recently had two new members join the team that Mattia works with. That made us six developers. We always pair program, so that makes three pairs. When they joined the WIP (work in progress) limit in development was set to two, and we purposely left it at two to encourage swarming - we love multiple pairs to work on the same story to get it done faster, whilst also increasing knowledge sharing and removing dependencies from individuals.

The new developers however didn't seem to like this heavily-collaborative style and wanted to increase the WIP limit to three, so that they'd be "free" to work on a separate story. We thought this would be a mistake, but realized that unless we let them experience the impact of the change they wouldn't fully understand the power of a low WIP limit.

So we agreed on increasing the WIP, and guess what? Our throughput went down. We were delivering less stories. Not only that: the lead time went up. Having learnt an important lesson, the new developers were now convinced and we went back to a WIP limit of two.

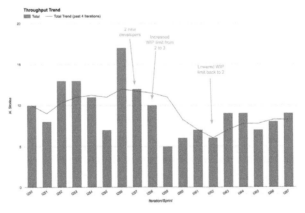

Real-life experiment: we used throughput to prove that increasing WIP was a bad idea

It's interesting to notice that even after setting the WIP limit back to two our throughput took a while to return to the original values. That's because of the cost of onboarding two new developers. Everyone in the team was going slower to take time to explain things to them and to let them learn.

Metrics that generate questions, not just answers

Mattia's team was having one of their (usually) monthly metrics review meetings. Because of a very busy period more than two months had passed since the previous metrics review meeting.

We noticed a significant drop in the throughput trend, which initially surprised us. "What could possibly be causing this drop? Does this confirm our feelings?" we asked. As soon as we started discussing it we realized that actually most people were feeling like it hadn't been such a productive period - being very busy didn't result in more work done, and actually some people admitted to having felt frustrated and having lost some motivation as a result.

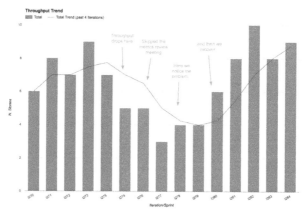

Real-life experiment: drop in throughput generated questions and surfaced important problems

Thanks to our metrics, a problem that was lying below the surface and intoxicating the team had become apparent, giving us a chance to fix it! We had a few retrospectives about these problems, identified some of the root causes and managed to recover, even increasing our throughput, when compared to before the busy period.

Remember, good metrics should surprise you every now and then. It's important that they generate questions as well, not just answers!

1.6 Use throughput for long term predictions

"How long will it take?"

"When are we going to complete this project/functionality/group of stories?"

Suppose we're analyzing a new project or a new piece of functionality and we need to answer the perennial questions "How long will it take? When will it be ready?".

If a simple back-of-the-envelope answer is enough for us then we can use the throughput average to make a simple linear projection, in a burn-up chart style.

Bear in mind that using an average for making plans or predictions can be very dangerous (the so called "Flaw of Averages": Plans based on average assumptions are wrong on average (Savage2012)). But if all you're trying to do is get a rough idea of duration then this could be enough.

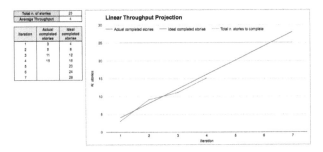

Simple linear projection using the throughput average

This approach assumes that you know how many stories you will need to complete, which is not always easy at the beginning of a project. There are some techniques to help you with this that we cover in chapter 6.

If you want more precision and, more importantly, if you need to manage the risks of the project, it's better to use throughput as an input to do probabilistic forecasting.

Probabilistic forecasting is explained in detail in the chapter on "Forecasting and planning". Briefly, it works by extracting random throughput samples from our historical data in order to simulate how much work the team will complete in each iteration. This is called a Montecarlo simulation.

The results are expressed as a range of probabilities: we're 85% confident that we can complete this group of stories in eight iterations, but we're only 50% confident that we complete it in six iterations.

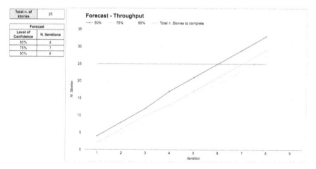

Use throughput for Probabilistic Forecasting and express results as levels of confidence

Real-life examples

Making predictions on the fly

Mattia and his team were discussing a new piece of functionality that would support the launch of a new product. Our business analyst (BA) presented the problem to the rest of the team, and we discussed possible solutions.

After agreeing on one solution, we broke it down into stories. Then the BA asked the perennial question "How long will this take?". We grabbed our throughput stats that we keep printed by our Kanban board, and quickly answered: "Well, we have ten stories here and our average throughput is three stories per iteration. So it looks like it would be roughly something around three to four iterations. Is this enough to answer your question? If not you can give us five minutes, and we'll plug these numbers into our forecasting spreadsheet to provide a more accurate forecast".

The BA replied "No don't worry, that's fine already. I just wanted to know if this would be ready for the launch of the new product in 6 iterations time. Sounds like we'll be fine". The meeting was over in less than an hour and we got started on that new functionality straight away.

1.7 Debunking some myths

"This only works if I have a lot of data"

You can start with a lot less data than you think:

- With 5 samples we are confident that the median will fall inside the range of those 5 samples, so that already gives us an idea about our timing and we can make some simple projections (see "rule of five (Vacanti2015)").
- With 11 samples we are confident that we know the whole range, as there is a 90% probability that every other sample will fall in that range (see the "German tank Problem (GermanTankProblem)"). Knowing the range of possible values drastically reduces uncertainty.

After that, each extra sample helps to further refine our precision. But a little data is enough to start. Also, when you have no data at all, even a small amount of data is a great improvement and can put you in the right ballpark.

"This only works if all stories have the same size"

This is a myth. It doesn't matter if stories have different size. There is only one size that we care about: "small enough". As long as we split stories as small as possible, then it doesn't matter if they have different size, and we simply use throughput to make predictions.

First, as discussed in the story points paragraph, the size of a story has little correlation with its lead time. So even if you think two stories have the same size, they can take a very different amount of time.

Second, the lead time for stories of the same work item type will follow a known distribution, regardless of their size.

"This is easy to cheat, I'll just create a lot of tiny stories"

Well, yes, but in this case that's actually a useful side effect. Smaller stories have a number of benefits, so encouraging that behavior is not a problem.

There is a point where stories become so small that the overhead of managing so many stories becomes greater than the benefits, but the team should be mature enough to realize when that's happening.

1.8 Public resources

In this book's public repository (https://github.com/SkeltonThatcher/ bizmetrics-book) you can find some useful resources for putting what we've talked about into practice:

- Link to a public Trello board that demonstrates how you can track information for your stories, including sprint when they were completed.
- Link to example spreadsheets that show how to analyze your data and generate all the throughput information that we've talked about.

1.9 Get started!

- Initially, don't drop whatever process you're currently following, but start collecting data and calculating your throughput. Simply count how many stories you have completed in each sprint.
- Use throughput at the next planning session to predict what you can achieve in the following sprint. Compare it with your usual velocity-based estimates.
- Use throughput for the next project to predict how long it will take. Compare it with your usual velocity-based estimates.
- Use throughput in the next retrospective to discuss if you're improving. Does it spark useful conversations?
- Compare your story points and throughput data and decide what gives you better predictability. If you decide to stop using story points, having data is going to be extremely helpful when it comes to convincing skeptical people.

1.10 Summary

Most teams have a need for predictability. We need to answer questions like "How much work can we complete? How long will it take? Are we nearly there yet?". Using the team's **throughput** we know how fast we're going and we can answer those questions.

By looking at our past throughput we can predict with confidence how much work we can complete in an iteration. The team becomes more predictable, making planning a lot easier.

Keeping an eye on throughput we can validate our experiments and decide whether the team is improving or not.

2. Lead Time: How long will this take?

Key points

- **Lead time is the amount of time that the story takes** from start (commitment point) to finish (the last state where you have influence).
- Using the lead time distribution we can **forecast how long a story will take** and **set expectations** that we're confident with.
- Using the lead time distribution we can answer questions like "**How long will this story take? Is this going to be ready in time? When should we start it?**".
- The lead time scatterplot is invaluable for **finding ideas for improvement**.
- Data on lead time helps us **decide what to work on today**. We focus on the stories that have been in progress for longer, as this makes us more predictable.

This chapter focuses on **lead time**, the time that a piece of work takes to go from start to end of a process.

2.1 Why should I care about lead time?

As a **team member**: by using lead time you become more predictable and gain higher confidence in your forecasts. You also spend less time estimating, which leaves more time for other work.

It becomes easier to decide what to work on next. Asking for help is more straightforward, as problematic stories become visible for everyone.

The amount of urgent work reduces, as you have data to fight back the pressure to start urgent work immediately.

As a **Scrum master** or **coach**: You clarify where lead time starts and stops, making the team's process more explicit for everyone. It's easier to notice when a story is having problems, so you can act on it.

Facilitating planning meetings and estimation sessions becomes much simpler as you focus on the value of stories rather than their size.

Lead time helps you get ideas for improvements and check if the team is improving.

As a **product owner**: lead time gives you realistic expectations of what the team can and can't achieve. You start thinking in terms of "What are the chances that this story will take less than 5 days?" rather than relying on the team promising that "they'll try".

You can use the team's predictions to have good conversations with stakeholders and focus on the value of a story rather than its size.

As a **manager**: lead time helps teams make realistic promises. You'll have a lot more confidence in the team's forecasts, allowing you to reliably do budgeting, planning and hiring. You'll have data to check whether decisions like hiring a new developer will help the team or not.

As a **customer**: lead time helps the team make realistic promises. Disappointments become the exception rather than the norm. You can be confident that when the team says they'll get something done, they probably will. You have useful discussions about the risks of a piece of work and the impact of possible outcomes, rather than arguing about delays and deadlines.

2.2 What is lead time?

"How long is this story going to take? When is it going to be ready? When should we start this story? Are we improving?". We can answer all these questions using our lead time.

You've probably heard the term "lead time" before and you might already know that it represents the time it takes for a story to be completed. But what does that mean exactly? When is a story considered "completed", and when should we start measuring the lead time? When does the clock start ticking, and when does it stop?

What does lead time really mean?

There is a lot of confusion around which terminology to use: is it "lead time", "cycle time", "time in process", or something else? We are using the names that seem to be most common in the Kanban community: lead time is the time from start to end; time in process (TiP) or time in X is the time spent in a particular part of the process (e.g. time in development).

Be aware that different people use the same terms when meaning different things, and use different terms when meaning the same things. Most famously, Dan Vacanti calls it "cycle time" in his excellent book "Actionable agile metrics for predictability" (Vacanti2015). The bottom line is: make sure that when you talk about lead time you clarify what you mean.

When does it start/stop?

It's important that you discuss and agree with your team on when lead time should begin and end.

Our suggestion is to start with the definition from the Kanban community: lead time is the amount of time that the story spends in any state in your process between your commitment point and the last state where you have influence (AndersonCarmichael2016). Let's first explain what those terms mean.

Lead time on an example Kanban board

The commitment point

The point in your process where a piece of work changes from being an option (for example, sitting in a backlog) to being selected as something that the team will work on next is the **commitment point**. When a story passes the commitment point the team is promising that they'll work on it and, if sufficiently mature, they communicate a completion forecast to the stakeholders. It often goes like this: "We're now starting this story that we know you're interested in, and we're 85% confident that it will be completed in 10 days or less".

The commitment point is where a story starts being considered "in progress", and it's therefore where the lead time starts.

For example in Mattia's team, where they use Kanban, their commitment point is on a column called "Next" that has a WIP limit of 2: in this column they want to have at most the next 2 top priority stories to work on at any time. Whenever there is an empty slot the team has a conversation to decide what the next thing to do should be. Pulling a story into "Next" is their point of commitment: they are promising to work on that story, and they know what its lead time is likely going to be, based on their historical data.

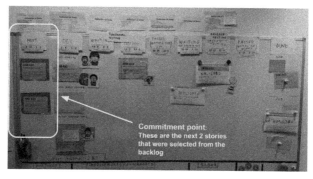

Commitment point on Mattia's board

Last state of influence

Lead time stops when the story reaches a state in the process where the team can't influence how long things will take from that point onwards.

An example could be a column called "Ready for Release": if the team can't release when the story is ready but instead has to wait for the next scheduled release, stories are going to accumulate in this column for some time. It doesn't make sense for the team to include time spent in this column in their lead time because they don't have control over it.

Note that they might, however, include it in what is sometimes called "customer lead time", to get a feel for what the customer experience is like. They might also want to measure how long the release step takes, so that they have data to prove that investing in continuous delivery is a good idea.

For example, in Mattia's team some stories can be released as soon as they are ready and some stories have to wait for the end of the sprint to be released in sync with other teams, depending on which codebases have changed. For the "release when ready" stories they calculate lead time as the time from "Next" to "Done".

Instead, for the "coupled release at end of sprint" stories the lead time is calculated from "Next" to "Waiting for cut", which is where ready stories wait until the end of the iteration.

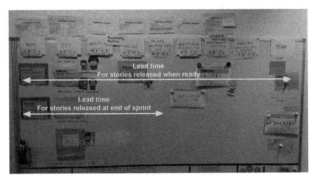

Lead time on Mattia's board

2.3 Measuring lead time

Measuring lead time is as simple as noting down the date when the story starts and finishes. If you use a physical board you could do this by writing down those dates on the cards, or you can extract them from your electronic tool if you use one. To calculate the lead time you simply have to count the days between the start and end date.

We measure lead time in days. When a story starts and finishes on the same day, we suggest you should count a lead time of 1 day for that story. It's tempting to try and be more precise and say that it took N hours, but in our experience we've never needed that level of precision - number of days is precise enough.

Why not measure only active time?

We measure lead time from start to finish, rather than just counting the active time on a story, because that's the only way to get any predictability.

In most development processes, the majority of time is spent in "queueing states" (e.g. backlog, ready for testing, waiting for release). A story that is actively worked on for just a few days can easily take weeks to reach the end of the process. If our lead time said "2 days", instead of the "2 weeks" elapsed time that the story actually took, it would be impossible to use that data to make predictions.

Split by work item type

We recommend analyzing the lead time of different work item types separately. Different work item types are usually characterized by whether:

- They follow a *different process* (for example, some user stories might not need to be tested, others might have to go through a special test environment).
- They follow the same process but at a consistently *different speed* (for example, work we do in our legacy codebase is consistently much slower than work on a greenfield app).

In the example from Mattia's team above, stories that get "released when ready" and stories that get "released at end of sprint" are two different work item types. Their lead time is going to be very different as they follow a different process, so they need to be analyzed separately. Mixing them up would give us data that wouldn't reflect either of those types of work.

Working days vs calendar days

When calculating lead time we recommend only counting working days. The reason is that if we counted weekends or bank holidays then the lead time would be higher, but the team would have no way of reducing that time. We've also heard of teams counting calendar days where people wouldn't start stories on a Friday to avoid being penalized with extra days on their lead time.

Be aware though that if you say to a stakeholder that "this will be ready in 10 days" they might think you mean calendar days. Make sure that you clarify what you mean.

2.4 Use lead time to make predictions

"How long will this story take?"

"Is this going to be ready in time?"

"When should we start this story?"

We can answer these questions with confidence by doing some analysis on the lead time of our past stories. Furthermore, using this data helps us have better conversations where we talk about value, discuss risks, and have arguments that are data-driven rather than being subjective opinions or best guesses.

Calculate lead time statistics to know your delivery time

It's useful to calculate some statistics on our lead time to get to know our delivery time. These stats can help us answer business questions like "How long will the next story take?" at a glance, allowing us to make predictions that we can be confident in. They also help us understand what range of data we're working with, and how much variability is in our process.

	Type 1: release at end of sprint	Type 2: release when ready
Average	5	5
Median	4	3
Min	0.5	0.5
Max	13	13.5
Mode	4	3
5% percentile	1	0.5
25% percentile	2	2
50% percentile	4	3
80% percentile	9	7
95% percentile	10.5	13

LEAD TIME STATS

Some useful stats on our lead time (for a particular work item type)

These stats are telling us a lot of useful information:

- Average and median: how long does a story take, on average? We prefer to look at the median rather than the average as the former is less impacted by outliers, thus providing a more accurate representation of reality. As statisticians say, "When Bill Gates walks into a bar on average everyone at the bar becomes a billionaire" (Ma2011).
- Min and max: what's the minimum lead time for a story? And the maximum? They help us understand the range of values that we're working with.
- Mode: what's the most common value of lead time? This is the number that occurs most often, so it's the one that people usually remember off the top of their head when asked "how long will this take?" (unless of course they have better data at hand).
- Probability percentiles: what's the likelihood that a story is going to be completed in this many days or less? Alternatively, you can read it as "how many days should we allow for a story to be completed, in order to be $X\%$ confident?".

Mathematically speaking, a percentile tells us that, when looking at 100 samples, N samples fall below a particular number. For example, in the picture above, the 80th percentile tells us that if we had a data sample of 100 stories of type 1, 80 of those stories would take 9 days or less. Percentiles are very useful for making predictions and setting expectations.

Lead time distribution

The lead time distribution is the probability distribution of the lead time. It tells us the likelihood of a story taking a particular number of days.

Knowing the lead time distribution is particularly useful for making predictions and setting realistic expectations with our stakeholders. If you've ever had predictability problems and are trying to improve, then this is the metric that you want to observe.

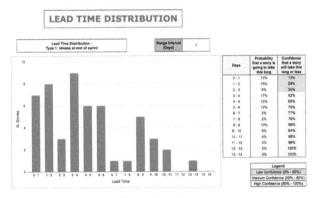

Lead time distribution

To calculate the lead time distribution you count the frequency of each lead time in your historical data. On the x-axis we have all the possible values of lead time that have occurred in the past, and on the y-axis the number of stories that have had that particular lead time. It's a histogram, which is why some people refer to this as just "lead time histogram".

In the example above, 7 stories have taken 1 day, 8 stories have taken 2 days, 3 stories have taken 3 days, and so on.

In the same picture on the right we also show some useful data to make it easier to interpret this information. We show:

- The probability of each individual lead time value, which represents the probability that a story is going to take *exactly that long*. For example, 13% of stories take 1 day, 15% of stories take 2 days, etc. This is calculated as the *number of stories with that lead time* over the *total number of stories*.
- The confidence that a story is going to be completed in *this many days or less*. For example, we are 52% confident that a story won't take longer than 4 days, and 88% confident that it won't take longer than 9 days. This is calculated as the *cumulative sum of the individual probabilities* (which is equivalent to calculating percentiles).

Making predictions

With this information, when someone asks "How long will this story take?" we can reply "We are 80% confident that it won't take longer than 9 days. It might take just 4, but there is only a 50% chance so I wouldn't make any promises on that. In the worst case scenario we might need 10 days, but there's only 10% chance that this might happen".

We are setting expectations that both the team and the stakeholders can be confident with. It's tempting to simply pick a number with a high confidence and answer "9 days", but we much prefer being clear with our stakeholders and explain that lead time is not a silver bullet for making predictions but it's rather a probability distribution. The only realistic answer is a range of possibilities.

There might be times when stakeholders insistently demand a single number. Even in this case we can still use our data by deciding how much risk we are willing to take on and provide a number that we are confident in. Be aware though that in doing so we lose one of the most important benefits of using metrics: having better, more informed conversations.

Focus on value and risk to trigger meaningful conversations with stakeholders

Having data changes the way we make decisions about the future: we think about how valuable a story is, the risks and impact of not having it ready in time, and whether that's acceptable.

Imagine that a stakeholder comes to us and says "We would really like this story done by the end of the month, to be included in the next release. Can you do it?". Is this feasible, or should we say "no"? If we said "yes", when should we start the story to have it done in time? Have we got enough time? Using the data at our disposal we know that we have 80% confidence that the story can be completed in time if we start it 9 days before the deadline. If instead we start it just 4 days before the deadline, then there is only 50% chance that it can be done in time.

If this conversation happens at least 9 days before the release, then we can say "yes" and be confident that we can make it. But what if there are only 5 days left? Now it's up to us and the stakeholder to discuss how valuable it is for this story to be included in the release, and therefore how much risk we want to take on when making this decision. If the story doesn't make the release, what's the impact? If we say yes but we don't make it, what's the impact? We're not just saying "we'll do our best" only to then disappoint the stakeholder, we're actually having a conversation.

Having these conversations changes the way we plan and estimate work: instead of focusing on the cost of an activity and trying to estimate and plan with more precision, we focus on the value of the work and the uncertainty that is intrinsically associated with it. Forcing ourselves to consider the risks makes us think "I know that there is only a 10% chance that this story is going to take more than 10 days, but what if it does? What would the impact be? Can we afford to take on this risk?". Some stories might be really valuable, so we can invest in reducing the risks for those. For other stories, we might be happy to accept the risk that maybe they'll slip to the next release.

Use data to set correct service level expectations

Data can also help us define an SLA (service level agreement) for our team: how much risk of being outside our SLA is acceptable? If we want a low risk - for example if our SLA is binding to a contract - then we would choose a high level of confidence and pick our 90th or 95th percentile.

However, more often than not, what's really at stake is what Dan Vacanti calls a service level *expectation* (Vacanti2018) towards the stakeholders, rather than an explicit agreement. In this case we can probably pick a less conservative percentile, like 80th or 85th. We know that every now and then there will be stories that fall outside this level of expectation, but they will be great opportunities for having a conversation and learning from them.

But always keep in mind that different types of work follow different distributions, so you want to set different expectations for those.

Why story size doesn't matter

There is a common myth that all stories need to be the same size for lead time to work. This is, indeed, just a myth. The size of a story has little correlation with its lead time. Other factors that impact lead time much more include how many things we're trying to do at once (work in progress) and how long a story spends waiting in queueing states. Even when two stories have the same size, they can take a very different amount of time to go through your delivery process.

There is only one size that we care about: **small enough**. As long as we follow consistent criteria for breaking down stories, our lead time distribution will consistently show us that we are going to have many stories that complete quickly, some that complete a bit slower, and from time to time one that takes very long. But we know exactly the chance of this happening.

Why not just use the average

Another common mistake is using the average lead time to make predictions. If you look again at the example stats and distribution you'll see why: we know from the stats that the average lead time is 5 days, but we also know from the lead time distribution that only 63% stories are completed in 5 days. If we communicated "5 days" to set expectations, when we know that stories take up to 9 days in 80% of the cases, we would end up disappointing our stakeholders quite often.

You should never make any predictions based on averages, as explained by the "flaw of averages" (Savage2012). The average lead time can be very different from reality due to outliers. A single number hides away the fact that occasionally stories take up to 13 days. We want to focus on the ranges of possible outcomes instead.

Using a single number to estimate uncertain future outcomes consistently provides wrong results. This is especially true if you're trying to forecast how long multiple stories will take. We talk about how to do it in chapter 5.

2.5 Use lead time to answer business questions

The data we have about our lead time helps us answer all sorts of business questions:

- **"How long is this story going to take?"** or the similar question "What's the estimated completion date for this story?". By looking at your stats and lead time distribution, you can answer with a set of possible outcomes accompanied by their particular likelihood. For example: "We are 80% confident that it won't take longer than 9 days. It might take just 4 days, but there is only a 50% chance so I wouldn't make any promises on that. In the worst case scenario we might need 10 days or more, but there's only 10% chance that this might happen". Some people won't like this answer and will insist on a single number. The first step in this case is always trying to explain what your answer means, but if they are not willing to listen you might need to decide how much risk to take on and give out a number that you are confident with. Beware though that in doing so we lose one of the most important benefits of metrics: having better conversations.

- **"Is this going to be ready on time?"** or the variation "Will it be ready by *X* date?". What does your data say about the chance of completing a story in the amount of days between now and date *X*? You should answer the question by explaining the likelihood that the story will be ready on time, and then discuss the impact of being wrong.

  Here is a real-life example: Mattia's team had 5 days left before providing a new version of their software for the next release. The next top priority story in the backlog was a new feature for an application used in call centers when customers phone up because they have problems with their broadband. The new feature was to show some extra information so people in the call centers could diagnose problems faster. The call center managers needed to know if this feature was going to be ready in the next release, because that would require organizing some training session for their staff (to teach them how to interpret the new information). The team looked at their data and knew that there was only 50% chance of completing a story of that type in 5 days. They also knew that this was going to be a risky story, because in order to test it they needed to use some real network devices, which have a tendency of being quite flaky. The team brought this information to the call center managers, and together they decided to move the story to the following release. If the team had simply replied "yes, we'll try our best to include it in the next release" and then failed, the managers would have trained people for nothing, causing confusion and more problems. Plus they would have had to repeat the training again later on, when the feature was actually ready. In this example, data showed the cost of being wrong was very high, so the managers preferred to take a decision that they could be confident about.

- **"When should we start this story that is needed by date X?"**, "When is the best moment to pull this story in?", "What sprint should we plan this in?". If the story is needed by a particular date, look at your lead time distribution and see how long, with a high level of confidence, the story might take to deliver. The last responsible moment to start it is that many days before date X. It's safe to wait until then, leaving other options open, because we're confident that we can complete it.

Another real-life example: Mattia's team needed to configure a new test environment for another team to perform some integration testing. They didn't need this environment for another 6 weeks. The team's data said that they could complete a story of that type in 9 days with almost 90% confidence. Therefore, the team stayed focused on what they were working on at the time, avoiding disruption. They waited until a couple of weeks before the environment was needed to start configuring it.

- **"Do this! We absolutely need it by X!"**. This is not really a question, but a rather common scene when someone is panicking and demands that something gets done. Having data really helps in this situation, because we can answer either "sure, we are confident we can get it done on time, no need to panic" or "based on our historical data, there is very little chance of completing what you're requesting in time, even if we stop everything else. What options have we got?". It's very human in a scenario like this to say yes under pressure - our data helps us stay subjective and avoid promising the impossible.

2.6 Use lead time for continuous improvement

"Where can we improve?"

"Are we improving?"

Lead time tells us how long stories will take, but we can use it for much more than just making predictions. Lead time can help us drive our continuous improvement process. Using our metrics we can find problems to talk about, get ideas for areas where we can improve, and then validate whether or not the changes that we introduce are helping.

Lead time scatterplot: where can we improve?

The lead time scatterplot shows, for each completed story, when that story was completed and how long it took.

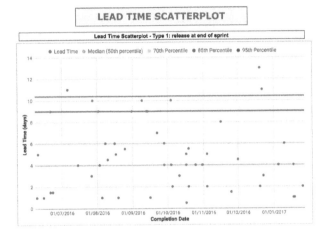

A lead time scatterplot

Each single dot represents one story. On the x-axis is the date when the story was completed. On the y-axis is the lead time of the story.

The horizontal colored lines represent various percentiles: they show us the percentage of stories that fall below that line, which are stories that have taken that long or less. In the example picture, the median tells us that 50% of the stories have taken 4 days or less, the 70th percentile tells that us that 70% of the stories have taken 6 days or less, and so on. In a similar way to how we use the lead time distribution, these percentiles give us the confidence of completing a story in a given number of days. For example, we are 85% confident of completing any story in 9 days or less.

As a metric, the lead time scatterplot has been getting more and more popular because there are a number of very useful things that you can read in it. We are going to focus on how to use it to answer one question in particular: where can we improve?

Finding ideas for improvement

To highlight the answer to the question "where can we improve?", we are going to re-draw the scatterplot with some light modifications:

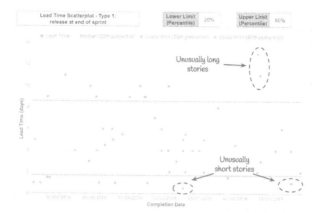

Scatterplot for finding ideas for improvement, highlighting interesting things to talk about

The focus is now on the two orange lines that represent our "limits". The lower limit tells us that only 20% of the stories take less than 2 days. The upper limit tells us that 80% of the stories take 8.5 days or less, which means that only 20% of the stories take longer than 8.5 days.

Stories that fall outside those limits are "unusual" - they are interesting stories that, for some reason, have taken either an unusually long or unusually short amount of time to be completed.

We want to talk about the long stories: what caused them to take so long? What problems in our process are highlighted by this? What can we avoid, or improve on?

But we also want to talk about the short stories: What happened for them to complete so quickly? What did we do well that we can repeat, or do more of?

Each story that falls outside the limits is an opportunity for improvement and we try and learn from it. Sometimes we understand it was just an isolated case. Other times we uncover a potential problem and we generate ideas to avoid the problem repeating in the future.

The lead time scatterplot is invaluable for finding ideas for improvement. It's a great metric to bring into a retrospective. It offers limitless food for thought and is fantastic for driving our continuous improvement process.

Which limits should we use?

We recommend using some percentiles as your upper and lower limits. Which particular values to use depends on your context, there is no hard rule. To help you choose, keep in mind that the purpose of this scatterplot is to generate interesting conversations: choose your limits so that you spark enough discussions and ideas for improvements.

In our example scatterplot we chose the 20th and 80th percentiles because they're generating a good amount of discussion for us. You could start with those and change over time as you improve.

What you shouldn't do is use the standard deviation to set the limits, which is an approach made popular by the control chart (Berardinelli) in Six Sigma. Doing that is only valid if you're dealing with a normal distribution, which as we've seen is not the case for software development processes.

Real-life example: what do these stories have in common?

During one of Mattia's team metrics review meetings the team was looking at the scatterplot for stories of type "release at end of sprint". They noticed that recently there had been a few stories that took unusually long:

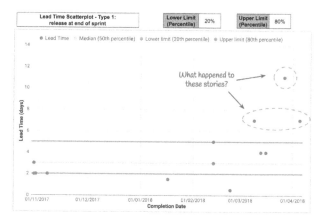

What happened with these highlighted stories?

As discussion went on, they realized all those stories had one thing in common: they were domain-heavy and involved complex business rules. For example, one was about implementing an algorithm for allocating broadband customers to the least utilized devices. Could this common theme be a sign that they were doing something wrong? Maybe they were waiting too long to clarify requirements? Maybe they didn't interact enough with domain experts?

When looking at metrics, you should try to avoid taking actions based on a single one. If there is a problem, you should see signs of that problem in other metrics as well. One metric can easily be gamed. Instead, a set of metrics doesn't lie.

Bearing that in mind, the team looked for either proof or disproof of their hypothesis in the scatterplot for stories of type "release when ready". Once again, most of the stories that were above the top limit involved deep domain knowledge. They were now pretty confident that there was a problem in this area.

The team spent the rest of the meeting discussing how to improve when working on domain-heavy stories and agreed on one important action: invest more time in the BDD process. Creating better acceptance tests from the beginning of the story should help clarify the business rules and make questions emerge as soon as possible.

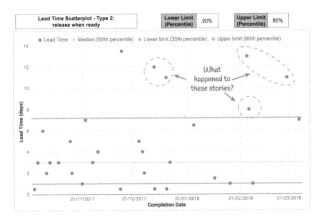

Do we see the same problem for stories of another type?

Lead time trend: are we improving?

To understand whether we're improving or not, it's useful to visualize the lead time trend: is our lead time growing or shrinking? Are we getting faster or slower?

Lead time trend

On the left side, we see the median of the lead time for all stories completed in a particular iteration.

On the right side, we see a rolling median over the last 4 iterations: each bar represents the median of the lead time for all stories completed in that iteration and the previous three. This has the effect of highlighting the trend for the lead time: observing whether the bars are getting taller or shorter we can understand if we're getting slower or faster.

We also show two lines for the trend of some percentiles, still calculated over the last 4 iterations. When the two lines are getting closer to each other it means that we are becoming more predictable, as the values of lead time are included in a smaller range. Percentile lines that are diverging instead mean that there is more spread in the possible lead time values, making the team more unpredictable.

The best use of the lead time trend is to validate the impact of the changes that we introduce in our process. When running an experiment, we look at this trend to answer the question "What effect is this change having on our lead time?".

Keep in mind though that it's only when multiple metrics show improvements that we can say we're improving. It's not enough for lead time alone to be getting shorter, as it might mean that we're gaming the metric (for example, by taking shortcuts that undermine quality).

Real-life example: multiple trends are no coincidence

In another metrics review meeting, the team was surprised to see that the trend for lead time was growing for a particular type of stories. They hadn't noticed any particular problems recently, so what was this trend about?

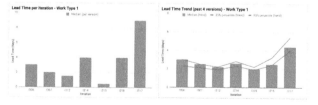

Lead trend was growing for stories of type 1

Once again, they looked for confirmation in trends for other types of work, to see if the same pattern was occurring:

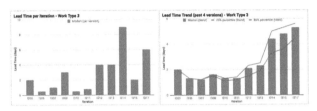

Lead trend was growing for stories of type 3

To their surprise, lead time was growing for all stories! On top of that, throughput (number of stories completed in one iteration) was going down, meaning the team was completing less stories per iteration.

This is an example of how metrics help you raise awareness for a problem that is not evident when executing the work. Good metrics don't just give answers, they generate interesting conversations.

The team started digging into what could have been the cause of the problem and realized that - after one team member had recently left on long term sickness - due to the rush and pressure of the current project they hadn't taken the time to review WIP limits. They were left with four developers but were still using the old WIP limit of 2. Because team members pair program full time, on most days there were two stories in progress.

Some people argued that there really wasn't a problem and that the limit should be kept at 2, while others advocated for reducing WIP limit to 1. In the end, the team decided to run an experiment: lower the WIP limit to 1 and revisit the decision at the following retrospective, in two weeks time.

A WIP limit of 1 proved challenging, but highlighted underlying problems in communication between the two pairs of developers. They started experimenting with a mixture of swarming (two separate pairs working on different tasks for the same story) and mob programming (all four developers working on the same task, taking turns to use a single keyboard).

After two weeks the results were promising: the stories completed in the last sprint had been faster than average, and team members felt like they were collaborating much better. The team decided to stick to a WIP limit of 1 until the fifth developer returned to work.

2.7 Use lead time to answer "What should we work on today?"

"Which stories are being problematic?"

"What should we focus on today?"

The metrics on lead time inform us on already completed stories. But we can use the same data to monitor the health of stories currently in progress and decide what work we should give priority to today.

Story health

For all stories that we are working on we look at how long ago they were started. We then use some percentiles, which we can easily calculate from our lead time distribution, to understand if the time that the story is taking is normal or if there's any reason for concern.

For example, if we used the same percentiles as the example data from the "lead time stats" section (50th percentile: 4 days; 80th: 9 days; 90th: 10 days), the story health would be calculated like this:

- 0% - 50%: when a story has been in progress for 4 days or less we consider it "green", everything is ok. Half the stories take this long, so it's a normal situation to be in.

- 50% - 80%: between 4 and 9 days the story becomes "yellow" because it's beginning to take longer than usual. We ask ourselves "Is everything ok? Are there impediments? Is it worth swarming on this and have another pair of people to help out?".
- 80% - 90%: from 9 to 10 days the story is "red". It's uncommon for a story to take this long. Maybe there is a problem that needs escalating? Maybe the scope was too big and we should break some parts out into a new story? We give priority to solving these problems and finishing these stories.
- 90% - 100%: After 10 days the story is "black", we consider it an emergency that should be solved with the highest priority.

You can visualize the story health both physically, putting tags on the cards on a physical board, and electronically, by color-coding the stories on your electronic boards.

In his team, Mattia makes sure that this information is up to date before the standup, so that they can decide what to work on based on it.

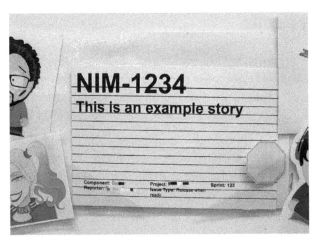

Physical story health - story with a yellow tag on our board

What should we work on? The older stories

We can look at the so-called "age of the work in progress" (Novkov2017) to decide the health of a story. The older a story is, the less healthy it is, and the more it becomes a priority to complete.

At standup meetings we can use this information to decide what to work on, giving priority to completing stories that have been in progress for longer.

By focusing on completing older items we become more predictable, as we keep the range of lead times under control. Shorter stories will lower the percentiles, which means that the team can promise that a story will be completed in fewer days with high confidence still.

Stories are also more likely to respect their forecasts and service level expectations, as the team proactively detects when there is a risk of overrun and then takes action to prevent further delays.

Statistically speaking, there will always be times when stories become red (more specifically, there's a 20% chance for each story) or black (10% chance). It's important to remind the team that this is not something to beat themselves up about, but rather something to celebrate as an opportunity to have a conversation and learn.

Disney stations

You know when you go to a theme park like Disneyland, you're queueing for a ride and you see those signs that say "Your queue from this point will take about 30 minutes"? And then as you get closer they say "20 minutes", "10 minutes", etc.? We use the same concept on our board to visualize a prediction of how long a story is going to take to reach the "done" state from where they are. We call this "Disney stations".

On top of each column on our physical board we print and stick a little piece of paper like the one in the picture. It shows:

- On the left, how long the story is likely to stay in this particular column of the board.
- On the right, how long the story is likely going to take to get from this particular column to "Done".

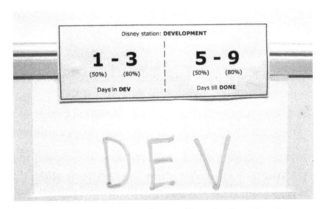

Disney stations: how long will it take from here?

For example the picture above, taken from Mattia's team, shows the Disney station that they have on top of their "Development" column. It shows that when one of their stories is in development, it's going to take 5 to 9 days to get it to "Done", and it's likely going to stay in the "Development " column for 1 to 3 days.

The two numbers in each section are the 50th and 80th percentile and represent different levels of confidence. Mattia's team is 50% confident that a story would stay in development for only one day, and 80% confident that it won't stay for longer than three days.

These numbers are calculated by looking at how long stories have stayed in each state of our process for, and how long they've taken to reach done from that particular state. On a regular basis (typically monthly) we re-calculate these values and print a new version of the Disney station, to replace the existing one.

Our Disney stations act as a real-time forecast for people that want to know how much longer it will be before the story they are interested in is finished.

Avoiding expedites

You should do everything you can to avoid expediting items (unexpectedly giving them higher priority over the work that is already in progress), because every time you expedite something you are actively contributing to reducing your predictability.

When you expedite a story, you drop what is currently in progress to work on it. By the time you resume work and complete paused stories, they end up with a longer lead time than normal. This in turn makes your range of lead times larger, which drives up the percentiles and increases your forecasts.

We should use our data to avoid expedites as much as possible. When someone asks us for something urgent, we can say: "We know with high confidence that few stories take longer than 6 days, and even fewer take longer than 8. We have a story that has been in progress for 5 days already, so it's very likely that it will complete today or tomorrow. So, if you just wait a day, we can work on this urgent task without having to impact the rest of the work".

Another effective way for minimizing the effects of expedites is to build "slack" into your process (Brodzinski2012a): keep some spare capacity reserved for emergencies or unexpected things. A practical example is using a work-in-progress limit that is lower than the number of people in the team. On a normal day the people that are not assigned to any regular story can work to help others, but they are ready to jump on emergencies when necessary.

And if you really can't avoid working on expedites, make sure you charge a premium price for it, to compensate for the problems that they cause. Using our Disney analogy again, it's a bit like when you pay extra to skip the queue for some rides: if the skip-queue service was too cheap everyone would use it, therefore making it useless.

2.8 Debunking some myths

"I don't care what you think, the data says so!"

The purpose of metrics is not beating people with them, but rather generating better conversations. Don't make the mistake of blindly following whatever you're reading in your data. Metrics are meant to inform your decisions, not override them. And sometimes they can be wrong too.

If the data says something but the people in the team say something else, chances are that maybe there's a problem in the data, or you modelled something incorrectly. Go back and check. At the end of the day, these are the people doing the job, no one knows it better than them.

"I can't do this because my boss wants a number"

There will be people that won't like hearing you talk about probabilities and possible ranges of delivery dates. They might say things like "What's all this math about? Just give me a number!". We have personally dealt with similar situations ourselves.

If you really can't convince them to listen then you can pick the result according to the level of confidence that you want to have (for example which lead time would give us at least 80% confidence?), and communicate that result to them. Be aware though that in doing so you are losing one of the biggest benefits of forecasting: discussing risks and options.

2.9 Public resources

Troy Magennis has a collection of free spreadsheets to cover many metrics needs. One about analyzing lead time in particular is "Throughput and Cycle Time Calculator".

Find them at http://bit.ly/SimResources

2.10 Get started!

- Initially, don't drop whatever process you're currently following, but **start writing down when a story starts and finishes and calculate its lead time.**
- Next time you're estimating a story, as well as doing your usual estimate, use the lead time distribution to **discuss the chance that the story will take longer, and what the impact would be if that happened.** Can you afford to take that risk on this story? What can you do to reduce it?
- As you get more confidence in your data, start using it to set expectations. **When asked to estimate, discuss the possible outcomes and their probability.**
- **Use the lead time scatterplot in a retrospective to find stories to talk about** that have been unusually long or short. What problems did you face? What went really well?
- Look at the trend for your lead time to **understand if you're improving or not.**
- At standups, be aware of how long a story has been in progress for. **Give priority to stories that are getting older** to improve your predictability.

2.11 Summary

Lead time is the amount of time that a story takes from start to finish. Observing the lead time distribution we know how long a story is likely going to take. This data helps us answer questions like "How long will this story take? When will it be ready?" with confidence and set expectations that stakeholders will trust.

Using metrics like lead time scatterplot and lead time trend we can easily find ideas for improvement and validate our experiments.

We also use our data on a daily basis to notice when a story is getting old, to focus our efforts on it. This makes the overall lead time shrink, which in turn makes the team more predictable.

3. Forecasting and Planning: When will it be done?

Key points

- The common way of estimating has several problems. Not only are traditional estimates very often **inaccurate**, they also usually convey a **false sense of certainty** and encourage behaviors like **big upfront planning** and **ignoring risks**
- When doing probabilistic forecasting we **use our historical data to simulate and forecast** what might happen in the future
- With a forecast you can **answer many business questions** like "**How long will it take?**" "**How much can we do?**" "**What should we do next?**". You can choose how confident you want to be with your decisions and how much risk to take on.
- **A forecast has many advantages** over traditional estimates. It's **more accurate** because it is based on data and measurements. It's expressed as a range of possible outcomes with their respective likelihood, to **communicate the uncertainty** about the future. **Risks are highlighted** early on.

This chapter focuses on **forecasting**, the prediction of work completion based on past throughput and number of items to complete.

3.1 Why should I care about forecasting?

As a **team member**: forecasting makes you more predictable, so you are able to make more realistic commitments. You have data to fight the pressure to say yes to unreasonable requests.

With forecasting you can concentrate on the problem and its solution without having to worry about putting a number on it. Also, forecasting is usually a lot faster than estimating, which leaves more time for other work.

Finally, meetings will become more efficient as decisions become less based on personal opinions and more on data.

As a **Scrum master** or **coach**: facilitating planning and estimation sessions becomes much easier as you focus on the content of the stories rather than their size.

You can highlight risks from the very beginning, and you can notice early on if you're not on track. You can also simulate what will be the effect of management's proposals on the team and convince them it's a bad idea, backed by data.

Finally, forecasting makes you more predictable, improving the relationship between the team, the managers, and the customers.

As a **product owner**: forecasting gives you realistic expectations of what the team can and can't achieve. You start thinking in terms of "What are the chances that we can complete this feature by this date?" rather than relying on the team promising that "they'll try".

As a **manager**: forecasting helps teams make realistic promises. You'll have a lot more confidence in the team's forecasts, allowing you to reliably do budgeting, planning and hiring. You'll have data to play what-if scenarios.

As a **customer**: forecasting gives the team the ability to make you realistic promises. Disappointments become the exception rather than the norm. You can be confident that when the team says they'll get something done, they probably will. When a problem arises you'll know as early as possible so that you can take action.

3.2 The common way of estimating

Quite often the very first thing that the team is asked, moments after hearing the requirements, is "How long will it take? Can I have an estimate?".

The common approach for an agile team to answer these questions is along these lines:

- **Analyze the requirements** with the help of the stakeholder or the PO and break them down into stories, forming a backlog.
- **Estimate each story in the backlog** using story points (or occasionally t-shirt sizes or ideal days).
- Use the team **velocity to convert story points into days** or sprints and find out how long it will take to finish the work. This could be the actual past velocity for an existing team, or an ideal velocity for a new team.
- **Pad the estimate with a "buffer"** to allow for emergencies or unexpected work. Sometimes a fixed amount of time is added, other times the whole estimate gets doubled or even tripled!
- Give the estimate to the client, expressed in terms of **cost and delivery date**. An agile team usually delivers work in an incremental way but it's still very common to have to declare the date when the whole project will be done.

Then over the course of the project, **the team checks if it's on track according to their actual velocity** and predicts when all items in the backlog are going to be completed, often using burn-down or burn-up charts.

Problems with this approach

This approach is without doubt a big step forward compared to the old waterfall days and teams have the best intentions when using it. Nevertheless, it usually leads to a few problems that shouldn't be overlooked.

False sense of certainty

The result is typically expressed as a single number or a precise date, with the effect of hiding the uncertainty that is intrinsic to any estimate. For example, "it will take 60 days" or "we'll finish on February 15th". The result might come with a caveat but people invariably fail to notice it.

Often it's the stakeholders who ask for a precise date because it makes them feel safe and they can use that date as a commitment. This date ends up in a plan and becomes a deadline that encourages a lot of unhealthy behaviors: changes are discouraged, discovery stops, collaboration is reduced, the team starts taking shortcuts when the deadline approaches, etc.

Inaccurate translation

As explained in the chapter on throughput, for most teams there is very little correlation between story points and lead time. Furthermore, velocity is one of the easiest metrics to cheat, especially when we're under pressure - without even noticing we end up doubling the points of each story to feel like we're going faster or to meet the expectations of a manager. This makes the translation from story points to days (or date) highly inaccurate, whilst still seeming precise. It's a recipe for disaster...

Unreliable when we're too busy

Queueing theory, from which Kanban takes a lot of inspiration, teaches us that the closer the utilization of a system gets to 100%, the more the lead time increases. In plain English, this means that activities take a long time when we're too busy (see Reinertsen2009 and Brodzinski2015b. The equation is known as Kingman Formula).

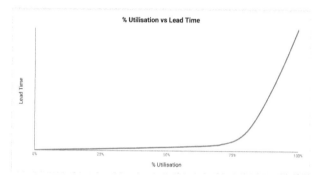

Lead time grows exponentially when utilization is high

High levels of utilization are typical in teams that don't limit their work in progress: whenever an activity is blocked they start a new one because they feel the need to be always busy, ending up with lots of half-done stories but nothing complete. In this scenario, estimates that are based on the size of an activity make no sense, since even the simplest of tasks is going to take quite long.

Troy Magennis (see forecasting experts) makes the perfect analogy to describe this problem: when a road is clogged up with traffic, then the bicycle, the Ferrari and the truck all move at the same speed - very slowly (Magennis2015).

Biased towards active time

When estimating, people tend to only think of the actual hands-on work time. However, in most development processes, stories spend the majority of their time in "queueing" states where no one is actively working on them (for example, in the backlog, ready for testing, waiting for release, and other wait states). Because of this, a story with an estimate of a few days can easily take weeks to reach the end of the process.

By the way, this is another reason why the size of the story doesn't matter much: size only impacts hands-on time, which ends up being a small fraction of the total lead time.

Biased towards expectations

Often we already have an idea of what the stakeholders want to hear. This ends up influencing our decisions to try and match the result that we believe will be accepted by the client. If the client doesn't accept then we say that our estimate is "wrong" and we review it. The estimation process now becomes a negotiation process where one of the parties has to lose (and, more often than not, the loser is the dev team).

Fixed-scope, no discovery

Often the team is asked to estimate a particular solution instead of being presented with a problem to solve. The team estimates it, creates a backlog of stories, and the project is not considered done until all stories have been completed. It becomes a fixed-scope project with no space for discovering what might add more value for our users. If we don't complete everything we consider the project to be late, without thinking if we should stop once we've actually solved the problem. We are too scared of stopping the project because we've already invested so much in it (the so-called "sunk cost fallacy" McRaney2011).

Encourages up-front plans, risks are ignored

Plans are made based on the estimate and are quite often "set in stone". From then on there is a deadline that the team must absolutely respect, often for no other reason than at some point in the past a promise was made to deliver on that date. There is no incentive to talk about what might go wrong, and even when we do, those risks are often ignored.

When a risk turns into a real problem, often the delivery date remains unchanged, forcing the team to find other ways of delivering the same scope. Teams do their best to adapt to changes, but they're usually reactive rather than proactive.

Expensive, rarely reviewed

The estimation process is often expensive: teams have to put a lot of effort into assigning a number to each individual story. Over the course of the project the estimates are rarely reviewed and it's only towards the end that we realize we are going to miss the deadline. It might be too late by then to take adequate corrective actions.

3.3 How long will this feature/project take?

"Probabilistic forecasting" is a technique used in many industries to predict the likelihood of uncertain events. The forecast is expressed as a list of possible outcomes accompanied by the probability of that particular outcome to become reality.

A typical example is the weather forecast: on TV they might say "Tomorrow it is going to rain in London" (which is always very likely) but what they really mean is "After using multiple models to simulate tomorrow's weather conditions, 80% of the simulations predicted rain, 15% predicted clouds but no rain, and 5% predicted snow".

The weather, just like software development, is a complex system where cause-effect relationships are not predictable in advance and are only visible in retrospect. We can't predict deterministically what is going to happen. What we can do instead is use our historical data to simulate and forecast what might happen in the future. This is what we call **probabilistic forecasting**. To do so we use a statistical technique known as Monte Carlo simulation (or Monte Carlo method). Here's how it works, step by step.

The input we need

We need to provide two inputs: "number of stories to complete" and "past throughput".

Number of stories to complete: If you haven't done so already, break down your requirements into stories. Remember that the size of individual stories doesn't matter, as there is very little correlation between the size of a story and how long the story takes. The amount of work-in-progress and queues in your process have a much bigger impact on lead time than size.

Probabilistic Forecasting - Input

N. Identified Stories: How many stories have we already identified?
15

Bugs Ratio: How often do we find a bug? Express as "1 bug every X stories"
10

Discovery Ratio: How often do we discover new stories, or split existing ones? Express as "1 new story every X stories"
5

Total N. of Stories: Identified + bugs + discovery
20

Past Throughput	
Iteration	Throughput: how many stories were completed?
1	4
2	9
3	7
4	5
5	4
6	10
7	5
8	1
9	4
10	7
11	2
12	3

Input data for our forecast: n. stories & past throughput

So don't worry too much about having similarly sized stories. As long as your team is consistent in the criteria that they follow for breaking them down, that's fine. For example in Mattia's team they follow this rule: "stories should be as small as possible, but they should still cover a vertical slice of the system and add value for someone - that someone being a user, a stakeholder, or the team itself". Other teams simply ask "Does this feel too big?" or "Does it feel like we could get this done in a day?". The end result should be stories that you feel comfortable working with.

In the total number of stories you should also leave room for bugs that you will find, discovery (such as new things that you will discover along the way), and stories that you might decide to split further (more on this later in this chapter, in the section "Account for dark matter: discovery, split rate, bugs").

Past throughput: As explained in chapter 2, throughput is the number of stories that you complete in each sprint.

 Note that, even though in this chapter we refer to 'sprints' for simplicity, you can apply everything we're saying to any period of time, like a week or a month. You don't need to be using Scrum to benefit from the metrics and approaches described your book.

Running the simulation

1) Let's begin the first simulation. We **pick a random value from our past throughput**, to simulate that in the first sprint we are going to complete as many stories. In the picture below we are simulating that in the first sprint we are going to complete 3 stories.

Total N. Stories to complete	Simulation 1			
	Iteration / Sprint	N. completed stories - in that sprint	N. completed stories - in total	N. Remaining stories to complete
20	1	3	3	17

In the first simulation, we are going to complete 3 stories in the first sprint

It's important to use historical data for this rather than relying on made-up numbers or averages. Our historical data reflects how our process works, so we're automatically taking into account our usual level of work-in-progress, the queues in the process, problems that tend to happen, and so forth.

Using the average would be particularly troublesome: the average being a single number hides away all the ranges of possible outcomes, which is what you want to focus on instead. The result would be a single value that falls somewhere in the middle of all possibilities, ignoring all the other scenarios that might instead be more likely to happen.

When dealing with variability in metrics like throughput and number of stories, using averages would be misleading (Flaw of averages - Savage2012). The only safe way to do calculations with them is to use a Monte Carlo simulation.

2) Repeat to run through the second sprint: how many stories are we going to complete? **Pick a new random value from your past throughput.** In the picture we picked 5, giving us a total of 8 stories completed after 2 sprints.

Total N. Stories to complete	Simulation 1			
	Iteration / Sprint	N. completed stories - in that sprint	N. completed stories - in total	N. Remaining stories to complete
20	1	3	3	17
	2	5	8	12

Sprint 2 of the first simulation: 8 stories completed so far

3) Repeat until the number of completed stories arrives at or exceeds 20. In the example it took us 5 sprints to complete the 20 stories (we actually completed 21). The first simulation is finished.

Total N. Stories to complete	Simulation 1			
	Iteration / Sprint	N. completed stories - in that sprint	N. completed stories - in total	N. Remaining stories to complete
20	1	3	3	17
	2	5	8	12
	3	1	9	11
	4	7	16	4
	5	5	21	0

In the first simulation we will have completed the 20 stories after 5 sprints

4) Repeat the simulation thousands of times (a spreadsheet is perfectly good for this, see our public repository for an example). By repeating the simulation so many times we are generating each possible combination of results, so that we can observe statistically how many times a particular outcome can happen. There will be simulations that are particularly lucky where it only takes us a couple of sprints, and other simulations particularly unlucky where it takes seven or eight sprints. But with so many repetitions these outliers become a minority compared to the more likely scenarios.

Simulation 2				Simulation 3				Simulation N		
Iteration / Sprint	N. completed stories - in that sprint	N. completed stories - in total		Iteration / Sprint	N. completed stories - in that sprint	N. completed stories - in total		Iteration / Sprint	N. completed stories - in that sprint	N. completed stories - in total
1	1	1		1	1	1		1	3	3
2	7	8		2	9	10		2	5	8
3	5	13		3	2	12		3	9	17
4	9	22		4	2	14		4	7	24
				5	4	18				
				6	1	19				
				7	3	22				

Repeat the simulation thousands of times to generate a statistical sample

Presenting the results

5) Count the frequency of each result to find its likelihood. For example, if 1700 simulations out of 10000 resulted in 3 sprints, that means we have a 17% chance of taking 3 sprints to deliver all stories.

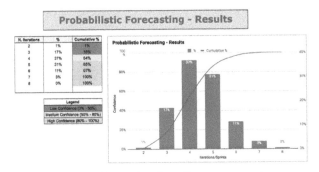

The results of our forecast

The picture shows that we have a 17% chance of taking 3 sprints, 37% chance of taking 4 sprints, etc. The column on the right is the cumulative sum of the individual probabilities: it tells us that we have 54% confidence in completing the 20 stories in 4 iterations or less, and 85% confidence in completing them in 5 iterations or less.

6) Present the results to the client as probabilities, like this: "We have about 50% chances of completing the work in 4 sprints, but we are 85% confident that it won't take longer than 5. In the worst-case scenario we might need 6, but there's only 15% chance that this might happen".

This style helps us to concentrate on managing risks: if the worst-case scenario was to happen and we took 6 sprints, what would be the impact? What can we do to reduce this risk? If the client insists on a specific deadline, then it's up to us as a team to decide how much risk we want to take on. If we wanted to feel confident we should go with 5 sprints. We could say 4, but then we'd have a 50% chance of being wrong and miss the deadline.

Don't be tempted to play the "padding game" and say 6 or 7 sprints: 6 sprints in this case has almost 100% confidence, which feels overly pessimistic. You'd be picking a level of confidence so high that you'd probably be requesting a lot of extra time that you won't need, damaging the trust between you and the client. We would rather be honest and say 5 sprints, which we are already very confident with. The only time we'd go for 6-7 sprints is if we were in a situation that required an extremely high level of confidence (e.g. dealing with a regulatory deadline).

Checking progress

Once we start working on the 20 stories we **use exactly the same technique to make sure we're on track**. At the end of each sprint we go back to step 1 and update our inputs: we add a new value to the past throughput, and we update the number of remaining stories. We then rerun the forecast to predict when we're going to finish the rest of the work. Running our forecast only takes a few seconds, so we rerun it every time we have new information. This way we can spot problems in our plans as early as possible and take action when there's still time to fix them.

3.4 Answering business questions with probabilistic forecasting

We can use forecasts to answer all sorts of business questions:

- **"How long** will it take to complete these 20 stories?". In our example above we were 85% confident that it will take 5 sprints or less to complete the 20 stories.
- **"What can we get done** in the next 2 sprints?". Run a Monte Carlo simulation to find how many stories you can confidently complete in 2 sprints. For example, "We are 80% confident that we can complete at least 7 stories in the next 2 sprints. Which stories would you like us to work on?"
- **"Will this feature be ready in time?".** Break down the feature into stories and run a forecast to find the likelihood of getting it done by a specific date. E.g. "With the current scope there is only a 60% chance of getting it done in time. But if we took out 3 stories, then our confidence would grow to 85%"
- **"Which feature should we work on next,** to be sure that it will be ready in time for the next release in 2 sprints? Feature A adds more value but comprises 11 stories. Feature B adds less value but only requires 3 stories". Run a simulation to see how long it would take you to complete the 11 stories in feature A. Say that the forecast predicts only a 40% chance of completion: in this case it's better to work on feature B next, which would definitely be ready in time for the next release.

Real-life example: choosing the next feature to work on

In Mattia's team they were doing some work on a product that is released once a month. This product is used by call center agents when customers phone in after experiencing problems with their broadband. There were two weeks left before the next release and the PO needed to choose between two features to work on.

The first feature was to change the way the software polled information from the network devices responsible for the internet connection. This would reduce stress on the devices and avoid having to restart them, ultimately saving money and improving quality of service for the customers. The total savings amount was hard to quantify but we expected to reduce costs by at least a few thousand pounds per month. This was a large feature, requiring 11 stories.

The second feature was to rearrange information on one of the web pages to make it easier to diagnose problems for customers with an old router model. This wasn't as valuable as the other feature, but still important for that type of customers. This feature only required 3 stories.

The PO wanted to know which feature the team members should work on for the next release, given that they only had two weeks left. They ran a simulation and predicted only a 40% chance of completing the first feature in time for the release, whereas they had high confidence that the second feature would make it in time.

With this information they decided to work on the second feature first, which made the release. The first feature was released the following month.

3.5 Forecasting tips & tricks: how to improve your forecasts

Forecasting number of stories

Sometimes it's not feasible to break down all the requirements upfront to find the total number of stories to complete, especially for big projects. In this case, what you can do is:

1. Split the requirements down into epics.
2. Pick a few random epics (5-7 should be enough) and break those down into stories. Randomness is important to guarantee that you're not biased towards bigger or smaller epics.

3. Run a Montecarlo simulation to forecast how many stories you will have in total: for each remaining epic, randomly pick the number of stories from one of the epics that you have broken down. Repeat a few thousands times to find the most likely result.

You can find an example spreadsheet for doing this exercise in our public repository.

Account for dark matter: discovery, split rate, bugs

One of the biggest mistakes that people do when forecasting is only taking into account the stories that they've already identified. The number of stories is always going to grow as the project progresses, because of what we call "dark matter" (Anderson2011):

- You discover **new work**, new requirements, or things that you hadn't thought about before.
- You **split existing stories**, because they're too big or they require preliminary work.
- You find **bugs** that you need to fix, leading to rework.

A rule-of-thumb guideline is to add 30% to 50% more stories than initially identified. If you break down your work and identify 10 stories, run your forecast assuming that you will actually have to do around 15 stories, because of things that you'll discover and bugs that will happen. You could even express the number of stories as a range from 10 to 15 and extract a random value from the range for each simulation.

An increase in number of stories anywhere between 30 and 50% is quite normal, with 100% being exceptional but not unseen. Ideally you would rely on historical data on how often you end up splitting initial stories and how often you encounter bugs in work already done.

The type of work matters

If different work item types follow a very different process you might want to run separate forecasts for each type, inputting the corresponding throughput for each particular type.

For example, say that you're forecasting a feature comprising 10 frontend stories and 5 backend stories. Your historical data tells you that in a sprint you consistently get 6-7 frontend stories done, and only 1-2 backend stories. That makes for a total throughput of 7-9 stories in a sprint.

If you tried to forecast all stories together using your total throughput, the result would look like you can almost certainly complete the 5 backend stories in one single sprint. After all, you usually complete at least 7 stories! But that would be completely wrong, since you know that you only get 1-2 backend stories done in a sprint.

You should instead run separate forecasts for the frontend and backend stories: the 10 frontend stories can be done in 2 sprints with high confidence, and the 5 backend stories will take 3-4 sprints. Assuming that you can work on frontend and backend in parallel, the feature will be ready in 4 sprints time.

Choosing the past throughput

When you select historical throughput to use as input for your forecast, the assumption is that the data was generated under roughly the same conditions as the project that you're forecasting. If the data was collected under significantly different conditions, such as a team double the size, or a project that used completely different technologies, then your forecast might be very wrong.

Choose input data that reflects your assumptions about the project you're starting. Then, as soon as the work starts, replace past data with your new actual data.

What-if scenarios for effective risk management

When forecasting, run what-if scenarios to identify which risks are more important to manage. For example, if the 3rd party system we depend on is not ready in time, we would need to implement a temporary solution. That would add 15 stories for us and result in a delay of 3 months. If this scenario were to happen, could we recover? Now we know the impact of this risk and we can take actions to manage it (for example, help the 3rd party provider plan their work, get regular updates from them, consider in-housing the system, etc.).

Trust the people

Don't make the mistake of blindly following whatever result the forecast comes back with. Forecasts are meant to inform your decisions, not override them. If the forecast says something but the people in the team say something else, chances are that there's a problem in the data, or you modelled something incorrectly. Go back and check. At the end of the day, these are the people doing the job, no one knows the reality better than them.

3.6 The benefits of forecasting

Probabilistic forecasting is not a silver bullet that magically solves all the problems with estimates. However, based on our experience, we can say that it at least mitigates them, whilst definitely making our life easier. Compared to the problems with estimates that we discussed earlier, a forecast helps:

- **Communicate uncertainty** about the future. While an estimate is usually expressed as a single number, a forecast is expressed as a range of possible outcomes with their respective likelihood.

- **Being more accurate** because it is based on data and measurements instead of people's guesses. The common problem of converting story points into effort and time is gone.
- **Take into account our typical utilization.** The historical data that we use automatically reflects our typical utilization level and how busy we usually are.
- **Take into account queues and past problems.** Since we're using historical data we are automatically accounting for queues in our process and any blocker or problem that happened in the past. With estimates we tend to be over-optimistic and only think about ideal scenarios. That rarely becomes reality.
- **Reduce personal bias.** Because a forecast is based on actual data, it's easier to stay grounded in reality and not give in to the pressure of always saying "yes" to what the client wants. When the data says that the desired delivery date is unrealistic we have objective evidence that we can use to discuss alternative options.
- **Keep flexible scope, allow for discovery.** The result of the forecast says that "in this X amount of time we can confidently complete N stories". However, it's not so important what the N stories are. The product owner, together with the team, can decide how they'd like to "spend" those N stories. The scope doesn't have to stay fixed to what was identified at the beginning of the project, allowing us to swap stories out for more valuable things that we discover along the way.
- **Encourage risk management and what-if scenarios.** With a forecast we think about the impact of something going wrong from the very beginning. We have a chance of playing what-if scenarios, which help us decide which risks are more important to manage.
- **Require less investment and discover problems early.** Creating a forecast is usually cheaper than an estimate. We still have to invest time into analyzing the requirements and breaking them down into stories, but that's all we have to

do for a forecast, whereas for an estimate you then have to come up with a number for each individual story. Forecasts are very easy to refresh every time we have new data (for example after every sprint). We constantly update our plans and have a chance of noticing early if we're going off-track.

The following table summarizes the advantages of forecasting over estimating:

ESTIMATE	FORECAST
False sense of certainty Expressed as single number or precise date	**Communicates uncertainty** Expressed as range of outcomes with likelihood
Inaccurate Low correlation between estimates and lead time	**More accurate** Based on data and measurements
Unreliable when we're too busy Even the simplest activities take forever	**Accounts for usual utilisation** Historical data reflects usual level of workload
Biased towards active time We only think about hands-on time, ignoring queues	**Accounts for queues and problems** Historical data includes queues and past problems
Biased towards expectations We give the result that the client wants to hear	**Reduces personal bias** Based on data instead of personal opinions
Fixed-scope, no discovery Fixates on a particular solution	**Flexible scope, allows discovery** Forecast N stories, not important which ones
Encourages upfront plans Often risks are ignored	**Encourages risk management** Highlights impact of risks
More expensive, no reviews Need to estimate each story; rarely reviewed	**Less expensive, reviewed often** Reviewed often with new data (e.g. every sprint)

Estimate vs Forecast

In summary, forecasting is better than estimating because forecasting:

- communicates uncertainty
- is more accurate
- accounts for usual utilization
- accounts for queues and problems
- reduces personal bias
- allows discovery
- encourages risk management
- is less expensive and reviewed often

These all lead to better business outcomes.

Case study: a better way of communicating delivery times

Robin Weston, Engineering Lead at BCG Digital Ventures, writes:

On a recent project our team had become frustrated with the constant demand for estimated delivery dates. The requested estimates were usually for a set of functionality, which in our world translated to a collection of several user stories. Even though we duly provided the requested estimates in good faith, our educated guesses were often held as fixed dates and any "late" delivery of functionality was met with a demand for explanation. As so often happens, we started to "pad" our estimates to give us some breathing room. We were disappointed in this outcome and resolved to introduce a better way of communicating expected delivery times to our stakeholders.

We had already been storing the start and completion dates of each individual story as they occurred in a Google Sheet. Therefore we decided to use this data to forecast a set of delivery dates along with a percentage chance of each delivery date being met. Firstly, we calculated the Takt time (Bakardzhiev2014) of each user story. We then augmented the Google Sheet with some custom Google Apps Script to forecast future completion dates for a user-provided set number of stories, i.e. to answer the question "When will feature X, which comprises 5 user stories, be complete?". The script would then use randomly selected Takt Times to run a Monte Carlo simulation of 10,000 possible future scenarios. From the outcomes of those scenarios we could select a range of delivery dates for the user story collection, alongside the probability of each delivery date being met.

Although our initial efforts were met with some skepticism, the

forecasting approach mentioned above worked well. Introducing the element of uncertainty that comes with any probability
based approach was a great starting point for any conversation
on delivery dates. Presenting our stakeholders with a range
of delivery date options allowed them to be more flexible in
their planning. The team members were also much happier as
we no longer had to worry about the political consequences of
estimations, while also benefiting from the extra time we had
previously spent in estimation sessions.

3.7 Debunking some myths

"This only works if I have a lot of data"

You can start with a lot less data than you think:

- With 5 samples we are confident that the median will fall
  inside the range of those 5 samples, which already gives us
  an idea about our timing (see "rule of five" (Vacanti2015)).
- With 11 samples we are confident that we know the whole
  range, as there is a 90% probability that every other sample
  will fall in that range (see the "German tank Problem" (GermanTankProblem)). Knowing the range of possible values
  drastically reduces uncertainty.

After that, each extra sample helps to further refine our precision.
But a little data is enough to start. If you have no data at all, you
can start with an ideal throughput (or that of a similar team) and
then replace it with real data as soon as possible.

"I can't do this because my boss wants a number"

There will be people that won't like hearing you talk about probabilities and possible ranges of delivery dates. They might say things like "What's all this math about? Just give me a number!". We have personally dealt with similar situations ourselves.

If you really can't convince them to listen then you can pick one of the possible results, according to the level of confidence that you want to have (for example how many sprints for at least 80% confidence level?), and communicate that result to them. Be aware though that in doing so you are losing one of the biggest benefits of forecasting: discussing risks and options.

"I need an expensive tool"

Although some software development lifecycle tools are starting to implement forecasting features, we still think that the best tool is a spreadsheet. When forecasting you want to experiment with different inputs, and nothing beats the simplicity of a spreadsheet for that.

"I can just use the average"

You should never make any predictions based on averages, as the "flaw of averages" explains (Savage2012). The average throughput can be very different from what happens in reality, and a prediction based on averages can therefore be highly inaccurate.

Moreover, the average being a single number hides away the ranges of possible outcomes, which is what you want to focus on instead. When dealing with variability in metrics like throughput and number of stories, using averages would be misleading. The only safe way to do calculations with them is to use a Monte Carlo simulation.

"If I don't estimate the team won't have shared understanding"

When forecasting, the team still gets the shared understanding from breaking down the work into stories together. You simply skip the step of putting a number on each individual story.

3.8 Public resources

In this book's public repository you can find example spreadsheets for the methods described in this chapter. It will help you get started with forecasting.

Also, Troy Magennis has a collection of free spreadsheets to cover many forecasting needs (http://bit.ly/SimResources).

3.9 Get started!

- Initially, don't drop whatever process you're currently following, but **start collecting data and calculating your throughput**. Simply count how many stories you complete in each sprint.
- Next time you're estimating a feature/project, as well as doing your usual estimate, **run a forecast and compare the results**.
- If the forecast is very different from the estimate, highlight that: "We estimated X, but based on our past performance **there is a chance that it might actually take Y**. What do you think?".
- As the work progresses, **keep updating your forecast** with new data, to see if you're on track or if alarms need to be raised instead.
- At the end of the feature/project, **compare** the time it took with both your estimate and forecast. Which one came closer? What are you going to do next time?

- **Reflect on the problems with estimates**: do they sound familiar? Do they happen in your team? Do you think forecasting could help you?

3.10 Summary

Most teams eventually get asked "How long will this epic/feature/project take?" or variations of it like "When will it be ready? What can we get done in the next N weeks? What should we do next? Is this going to be ready by date X?".

In this chapter we explained how to answer those questions with confidence using a technique called probabilistic forecasting: using our historical data to simulate and forecast what might happen in the future.

A forecast has many advantages over traditional estimates. It's more accurate because it is based on data and measurements. A forecast is expressed as a range of possible outcomes with their respective likelihood, to communicate uncertainty about the future. Finally, risks are highlighted early on.

4. Metrics for Flow: Where is the bottleneck?

Key points

- **Flow is the movement and delivery of customer value** through a process.
- **Good flow leads to many benefits** for the team, including **faster delivery, predictability, earlier feedback** and more **motivation**.
- We can observe the flow in our team and **spot any impediments** by using a **cumulative flow diagram** (CFD) and a **net flow** diagram.
- Work spends most of the time in "wait states", as a low flow efficiency shows. We can **vastly improve by focusing on reducing wait time**.
- Metrics like *release time* and *time in state* help us have data-driven conversations and **focus on improving the areas that will have the largest impact**.

This chapter focuses on **flow**, the movement and delivery of customer value through a process.

4.1 Why should I care about flow?

As a **team member**: good flow makes stories go through the process faster, so you'll get feedback sooner. Your work gets released sooner, which is much more motivating than having a long wait period until release. There are fewer blockers and impediments, so you're stuck less often. When there are problems, the rest of the team is ready to help you. You pick up new skills by collaborating with people in other roles and become more cross-functional.

As a **Scrum master or coach**: finding and removing impediments to flow drives the continuous improvement process for your team. To achieve better flow you encourage more collaboration between everyone in the team, which results in a better, happier, team.

As a **product owner**: a team with good flow delivers work faster and with high predictability, which builds trust between you and the stakeholders. Customers can see that we're releasing on a regular basis and that their requests are taking less time to be implemented, improving their satisfaction.

As a **manager**: you have data to find the areas for improvement that will have the biggest impact, giving you indications of where the process needs changing the most (for example, releasing more often).

As a **customer**: you can see that the team is releasing on a regular basis and that your requests are taking less time to be implemented. You have a chance to give them more feedback and see them acting on it very quickly.

4.2 What is flow

"Is work going through our process in a smooth and predictable way?"

"Are stories moving regularly on our board?"

"Are there impediments/blockers?"

Seeking good flow is one of the most effective ways for teams to pursue continuous improvement: **keep on identifying the most serious impediments to flow and strive to remove them** (Burrows2014). In this chapter we'll show you some metrics to do that.

Focus on good flow to achieve faster delivery and other benefits

Flow is the movement and delivery of customer value through a process (Vacanti2015). It's how consistently and continuously work moves through your process all the way to a customer. Typical questions that teams ask about flow are: "Are stories moving regularly on our board? Do they move independently or in big batches? Are there impediments/blockers? Is there a bottleneck, and if so, where is it?".

Bad flow usually looks like big piles of work waiting to be processed. At every step of the process there is a large queue of unfinished work. Any new story that is started is going to take forever to be finished, even though the story itself might be quite small. Stories spend a long time sitting in queues between activities - for example: waiting to be tested or waiting to be deployed. Releases are infrequent and contain many changes, which increases risks and leads to big ceremonies, large coordination cost and management overhead. The teams spend a long time in meetings to manage large backlogs, discuss big release plans, and plan large activities like projects and big features. Work is often blocked and as a response new work is started, only aggravating the flow problems. People are overworked and unmotivated. They feel that they are working hard but not achieving much.

Good flow looks like a continuous stream of value moving across our process and being delivered to our customers. Work moves steadily and in a predictable way, usually in small chunks rather than in big batches. The amount of work in progress is balanced to the capacity of the system. When there is good flow we don't see work piling up or waiting for a long time. Blockers and impediments are rare, and when they happen the team swarms to solve the problems quickly. People are not overworked. There is enough slack to handle emergencies and unpredictable events like bugs and urgent requests.

Good flow leads, first of all, to **faster delivery**. Common ways to improve flow are reducing wait time and doing things more frequently but in smaller batches (that is, more frequent, smaller releases). The effect is that our lead time becomes shorter - stories go through our process much quicker.

Secondly, good flow improves **predictability**. Stories spend less time waiting or blocked, which are usually the unpredictable parts of our process. As a result, the team becomes more predictable and can make better forecasts.

Third, the team gets **feedback earlier**. As we release more frequently and the lead time shortens, we're putting work in front of our customers more often, giving us a chance to get feedback sooner. And the more we do this, the more we validate whether we're on the right track.

Fourth, it increases **customer satisfaction**. Customers can see that we're releasing on a regular basis and that their requests are taking less time to be implemented. This builds trust and improves relationships with our customers. They can give us feedback more often, and because of our short lead time the feedback is turned into actions very quickly.

Fifth, focus on flow drives **continuous improvement**. By observing flow, we notice where there are impediments and have opportunities to improve.

Sixth, it creates **more collaboration**. Another common practice to improve flow is for people to be more cross-functional and swarm more often, which improves collaboration and results in a better, happier team.

Finally, people get **more motivation**. Seeing that things move quickly and knowing that something could go to production the next day gives people more sense of urgency and motivation. Receiving more feedback helps everyone feel that what they do is meaningful.

4.3 Use a cumulative flow diagram to answer: Is work flowing well?

"Is there a good flow in our process?"

"Where are the impediments to flow?"

A **cumulative flow diagram** (often called just **CFD**) is one of the best metrics to observe flow. A single diagram provides a lot of information - so much so that people new to it can sometimes find it a bit hard to read. But don't worry, in this section we'll show you how to generate one and the various ways of using it.

How to draw a CFD

Generating a CFD is actually quite simple: for each day over a period of time, plot the number of stories that are in each state of your process on that particular day.

On the x-axis is the date, and on the y-axis is number of stories there were in each state on that date. In the example, taken from one of Mattia's teams, on the example date there were 5 stories in "Done" state, 6 stories in "Release Testing", 2 stories in "Waiting for release testing", 5 stories in "Waiting for cut" (waiting to be included and deployed in the next release), 2 stories in "Development" and 1 story in "Next" (selected as the next top priority story to work on).

We are also showing when each iteration/sprint started (I251, I252, etc.).

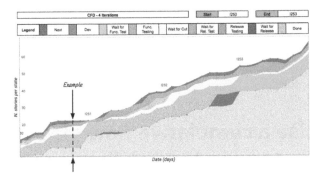

Example CFD: plot the number of stories in each state for each day

The CFD is all about observing the flow in our team. By looking at how the band for each state behaves we can learn a lot about how work is flowing through our process. A CFD for a team with good flow should show:

- The bottom band (e.g. "Done") growing steadily, meaning that stories are being completed on a regular basis. As the bottom band grows it pushes up the other ones.
- Thin bands, meaning that there isn't too much work in progress in that state. The only band that should keep growing is the bottom one (e.g. "Done"), where things are obviously going to accumulate.
- Bands that grow steadily and in parallel, meaning that work is moving to the next state at a regular, predictable, rate.

An "ideal" CFD, representing perfect flow with all these characteristics, would look like this:

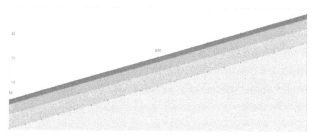

The ideal CFD has thin bands that grow in parallel

Practical tips for generating great CFDs

Collecting data

Collecting the data for a CFD can be as easy as simply counting the stories in each state. Be aware though that there will be times when you realize that you need to "fix" some old data: maybe you realize that you've made a mistake, or maybe a story was in the wrong state when you recorded the data. In our experience we found that, rather than counting, it's better to record the transitions between states: write down when a story moves from one state to another, and then you can easily fix this data when you've made a mistake and re-calculate how many stories were in each state each day using a script or a spreadsheet.

Consider different time periods for different actions

Depending on what we're using the CFD for, we might want to look at a longer or shorter time period. In our experience, this is a good rule of thumb:

- Up to a month (current sprint + previous sprint) for daily usage to proactively spot problems as soon as they appear and identifying patterns that are starting to emerge.

- Up to two months (past 4 sprints) for analyzing that period of time in a retrospective. It's long enough to see patterns, but not too long to have forgotten what happened.
- Up to 4-6 months (past 10-12 sprints) for seeing bigger trends in the team, especially comparing the flow now to a few months ago.

Start from zero

On the left-hand side of the CFD, at the beginning, always make the count for the bottom line (e.g. "Done") start from 0, regardless of how many stories were already "Done". When you're generating a CFD for the last two weeks, it's likely that two weeks ago there were already lots of stories in the "Done" state.

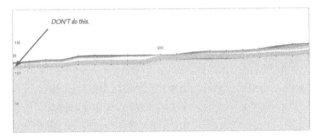

Don't do this! Always start the bottom line from 0

For the sake of the CFD you want to make your "Done" count start from 0, and only consider the *new* things that are moving to "Done". If the bottom line started already at a high number, the other bands would appear squished and it would be difficult to see anything useful.

Visualize all states, especially the queues, to improve flow

Sometimes it's useful to simplify a CFD to highlight a particular aspect of our process, or the effect of a particular problem. In general, though, we recommend showing all of our states, especially the ones which represent wait time. Stories in these states are not being actively worked on, they are just waiting for something to happen. For example, "Ready for testing", "Waiting for deploy". Managing these wait states is key to improving flow. visualize them on your CFD so that you can notice when queues are growing too much and work is waiting for too long.

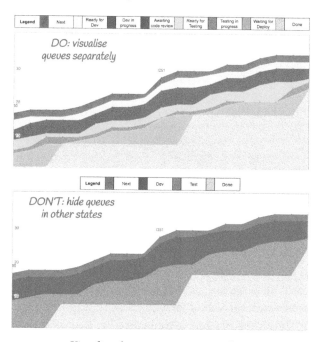

Visualize the queues to improve flow

Annotate additional information for context

We like annotating our CFD with additional information, for example when a particular iteration/sprint starts (e.g. I251, I252) or when important events happen (e.g. start of a new project, or a new person joins). This makes it easier to relate to the time period, and helps us understand what might have contributed to the patterns that we see. If you use a tool that doesn't let you add annotations, you can always print the CFD and write on it before you analyze it.

Focus on arrivals and departures

To observe and improve the flow in our team, the two key points of our process to pay attention to are where items enter the process ("arrival") and where items exit the process ("departure"). We can adapt the CFD to highlight these two points:

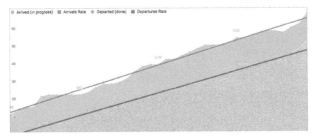

Arrivals and departures on a CFD

An item is considered to have **arrived when there is a commitment to it**, when it has been selected as one of the next things to work on. The commitment point is where a story starts being considered "in progress", and it's also where the lead time starts. In a Scrum team, this could be when a story is put into the "Sprint Backlog". In a Kanban team, it could be when a story is picked from the backlog into something like "Next" or "Selected for dev". Stories in other backlogs are usually still options, there is no commitment to them, therefore it's usually best not to include them in a CFD.

An item is considered to have **departed when it reaches the end of our process**, however that end is defined (e.g. "Done", "Released", etc.). All the items that have arrived and not yet departed are in progress.

For a good flow, we want the arrivals and departures to grow at the same rate. If arrivals grow faster than departures, it means we're starting more work than we're finishing, which leads to bottlenecks and stories taking longer. This is often the case in teams that don't limit their work in progress but start new activities every time they are waiting or blocked. On the other hand, if departures are faster than arrivals, it means that we're finishing work faster than we're starting it, so we risk ending up starving the team of work. We want to start and finish work at the same rate, to keep our process balanced (Vacanti2015).

Highlighting on the CFD the trend lines for both "arrival rate" and "departure rate" helps us keep them under control. Another great metric for this purpose is the Net Flow, which we are going to present later in this chapter.

Common CFD patterns and what to do about them

CFDs contain a lot of information and can sometimes be hard to read. This is a list of common patterns that you can spot - each representing a different problem - and what to do about them (Brodzinski2013).

CFD pattern: diverging lines

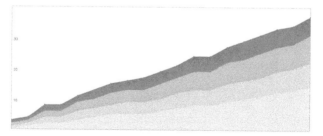

Diverging lines: we're starting more than we're finishing

Pattern: Lines are diverging, bands are growing.

Problem: We are starting more work than we are finishing. This pattern is common in teams that don't use WIP limits: whenever a story is waiting or blocked, people start a new one because they think they should always be busy. Over time the amount of work in progress keeps growing, and the lead time of stories becomes longer and longer.

Action: Introduce WIP limits. Focus on finishing stories that are already in progress before starting a new one. In practice, this means that sometimes people will have to help in steps of the delivery that are not their specialty (e.g. a developer doing some testing). Even at reduced effectiveness, that's still a better choice than starting new work. Plus, in the long term there's a clear benefit of people picking up new skills and becoming more versatile. Remind people that their job is not to write code, but to add value - there are many activities where they can add value that are not necessarily writing code.

CFD pattern: steps

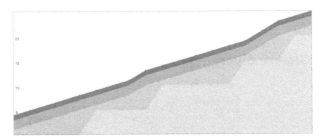

Steps: we're releasing infrequently in big batches

Pattern: Lines look like staircase steps.

Problem: Work is moving in big batches. A typical example, which manifests itself when the steps are in the bottom line like in the example above, is when releases are infrequent and contain many changes that have been batched together. Big releases like that have many disadvantages: they increase the risk that something will go wrong, they make diagnosing problems harder, they delay feedback, and if a rollback is needed everything is rolled back together.

Even when the stepped line is in another part of the process, it's still a problem. We're moving a batch of stories all in one go (e.g. multiple stories ready to test all at once), instead of getting feedback as early as possible on each individual story.

Action: Introduce WIP limits to reduce your batch size. If releases are batched, look into moving to continuous delivery. A common reason for delaying releases is when they are expensive (e.g. if they need to happen overnight). Look into these reasons and work to reduce this cost. If another part of the process is batched, introduce WIP limits to initially keep the batch size under control, and then incrementally reduce them to make your flow better and better.

CFD pattern: flat lines

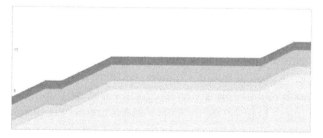

Flat lines: nothing is moving, there is no flow

Pattern: Lines are flat for several days in a row, or grow very little. A variation of this pattern is when a single line is flat.

Problem: There is no flow, stories are not moving on the board. Work stays in the same state for a long time. Flat lines are the worst enemy of flow because they mean that nothing is getting done, no value is being delivered to the customer.

Action: One of the first things to look at in this scenario is reducing the size of stories - find ways to split them into smaller chunks that can flow more rapidly through your process. It's also possible that work is not moving because of blockers or impediments (for example, a broken test environment), in which case you need to take actions to remove them (for example, finding the right people that can fix your test environment).

CFD pattern: bulge in a state

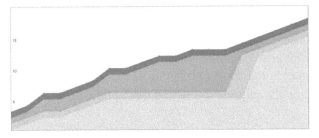

Bulge in one state: we have a bottleneck

Pattern: One band keeps growing and growing, creating a bulge.

Problem: There is too much work in progress in that state and it has become a bottleneck. It's not coping with the amount of work that is being pushed into it. Chances are there is little collaboration in the team and people are sticking to their responsibilities rather than going to help where their teammates are having problems.

Action: The immediate action should be to "stop the line" and clear the amount of work stuck at the bottleneck. Then, introduce WIP limits to prevent this from happening again so that you won't be able to push that much work into that state again without consciously breaking the limit. Training people to be more cross-functional (Yip2018) is also a good idea, so that they can help in the future when stories in that state are not moving. In general, investigate ways to improve collaboration between the members of the team.

CFD pattern: dropping line

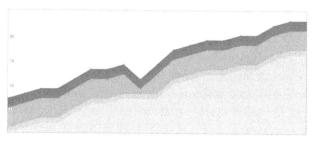

Multiple lines drop: stories were started and then thrown away

Pattern: Top lines are dropping.

Problem: A story that had already been started has been discarded and thrown out of our process. Maybe we realize we don't have all the information we need to proceed, or that it's not a high value piece of work to invest in now.

Action: Clarify your policies about when a story is ready to be started. How can you prevent this problem in the future and identify earlier that the story should be discarded? Which (missing) steps in our process are letting this happen?

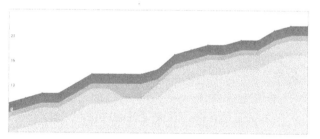

Single line drops: something moved back on the board

Pattern: Single line is dropping.

Problem: A story that had already progressed to a particular state is moving backwards on the board and going back to the previous state. The blue band increases because the story that had previously moved to green has gone back to blue, but no new work has been pulled in so the red line is not moving. A typical example is when teams find a problem while testing a story and they send it back to development.

Action: If the story had moved forward due to a genuine mistake, as in it was never meant to move to testing, then simply remove the timestamp of that transition from the story and redraw the CFD. If instead the team found an actual problem and wants to move the story backward, stop and reflect. Flow should only ever go in one direction. A better strategy is to leave the story where it is and mark it as blocked, while creating a new story (e.g. a bug) to fix the problem. The bug goes through the normal process until it reaches the blocked story, which can now be unblocked and progress again. We recommend that, while the story is blocked, you still make it count towards your WIP limit, so that you're encouraged to fix the problem.

CFD pattern: timebox

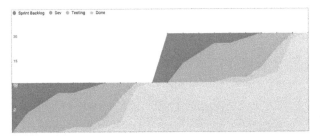

Typical CFD when using a timebox (e.g. sprint)

Pattern: Big jump at beginning of sprint, then all work gets done at the end of sprint.

Problem: This is the typical pattern of a team that uses Scrum and works in sprints. New stories for the next two weeks get moved to the sprint backlog. Soon, most of them go into development, and in the last few days of the sprint stories are tested and moved to

"Done". Work is moving in big batches of two weeks.

Action: Look into ways to make your cadences independent. Releasing and replenishing your queue of work don't necessarily need to happen in sequence. You could decide to pick the next stories from the backlog every three days, and still release every two weeks. Looking into some material on Kanban would give you ideas to improve your flow.

Little's Law on the CFD

An interesting characteristic of the CFD, one that you often see mentioned in articles and talks about it, is that you can use it to extract other metrics, in particular the ones about Little's Law (Lowe2014b).

Little's Law comes from queueing theory, which Kanban takes a lot of inspiration from, and states that $averageLeadTime = \frac{averageWIP}{averageThroughput}$ (Lowe2014a).

In simple terms it means that, on average, if you want stories to go faster (shorter lead time), you either have to increase the throughput (get more done) or reduce the amount of work in progress (work on fewer things at the same time). Little's Law is about averages, so it can't really be used for making predictions. However, knowing whether Little's Law is being respected in your process gives you a good indication of whether your process is balanced and therefore has good flow. If you want to know more about this, Dan Vacanti gives an excellent explanation in his book (Vacanti2015).

We can read all three of Little's Law metrics in the CFD:

- The vertical distance between the top and bottom line is the amount of work in progress (WIP), since that's how many stories were in those states of our process on any given day.

- The horizontal distance between the top and bottom line is the approximate average lead time of a story, since that's how long it has taken between some stories starting and some stories finishing (this lead time is an approximation because it's not necessarily the same stories that have started and finished).
- The slope of the bottom line represents the average throughput, since the steeper it is the more items we're completing. Similarly, the slope of the top line represents the average arrival rate, because the steeper it is the more items we're pulling into our process.

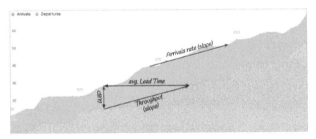

Little's Law metrics on a CFD

This is very useful for scenarios where the CFD is the only metric you have. However, we always recommend that, if you can, you avoid overloading a single chart with too much information and instead you generate multiple metrics that are easier to understand. If you have access to the data that was used to generate the CFD, it's very easy to calculate lead time, throughput and WIP separately and with more accuracy.

Validating experiments with the CFD

We can use the CFD to validate our experiments and check whether the changes that we introduced to our process are improving the flow.

Real-life example: drastically better flow after reducing the WIP limit

This is a real-life example from Mattia's team. The team had 6 developers, and since they pair program full time, that meant there were 3 pairs at any given moment. The team was struggling with flow: it felt like nothing was moving and every story was taking a long time to complete.

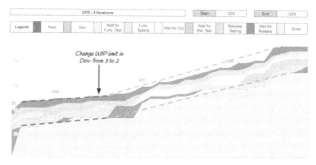

Real-life example: flow drastically improved when reducing WIP limits

This is clearly visible on the left side of the CFD, where the flow is almost flat. During a retrospective, the team talked about the fact that the current WIP limit of 3 in "development" was encouraging each pair to work on different things. So, they decided to run an experiment: for the following 2 weeks, reduce the WIP limit from 3 to 2. The extra pair was tasked with helping flow - on a normal day they would swarm with another pair and pick different tasks from the same story, and they would also deal with any emergency and interruptions like emails, meetings, etc. The change in flow, which you can see on the right side of the CFD, was so dramatic that everyone in the team was more than happy to consider the experiment a success and keep the WIP limit at 2.

4.4 Use net flow to answer: Is our process balanced?

"Do we start and finish stories at the same rate?"

Another good indicator of flow is looking at how many stories we start and finish in a particular period of time and checking if there's a balance. When we start and finish items at the same rate we are keeping the amount of work in our team balanced, and stories are flowing through our process at a steady and predictable pace.

Net flow is a great metric for visualizing this balance (or, rather often, unbalance). For each period of time (e.g. each week) it shows the difference between how many stories were started and finished, calculated as: $netFlow = numberOfStoriesCompleted - numberOfStoriesStarted$ (MagennisSpreadsheets).

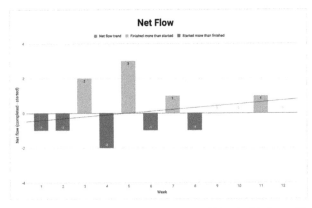

Net Flow: is the amount of stories started/finished balanced?

In this example, in week 1 the team started one more story than it finished. Same for week 2. The team caught up in week 3, when it finished two more stories than started that week. Weeks like 9 or 10, where the Net Flow is 0, are weeks where the team started and finished exactly the same number of stories.

We shouldn't always expect a perfect net flow of 0, it's perfectly normal to see some fluctuation each week. Which is why visualizing the trend is useful: if the trend is going down towards negative numbers it means that we're starting more work than we're finishing. This is going to lead to an increase in the amount of work in progress, which we know results in stories taking longer, receiving slower feedback, and showing lower quality.

Finishing more work than we're starting might not sound as bad, but it's nevertheless a dangerous situation to be in: it's uncovering a bottleneck upstream of our process, which is causing not enough work to come our way. If we don't do something about it the team might soon starve for more work.

4.5 Use flow efficiency to improve the end-to-end process

"How much potential improvement is there in our process?"

"Are we improving the whole process or only individual parts?"

"Are we focusing on end-to-end flow or on keeping people busy?"

In nearly all development processes, if you look at the steps that a piece of work goes through from when it's started to when it's completed, you'll find that work spends a lot of time not actually being worked on, but rather "waiting" for something to happen. At the team level, typical examples are stories sitting in the backlog, or a story waiting for a tester to be free, or waiting for the next scheduled release.

Flow efficiency is the percentage of time that work spends being actively worked on, as opposed to the total amount of time the work takes, including wait time.

Calculating flow efficiency

On our board we can easily highlight which states are considered to be "**touch time**", where stories are being worked on, and which states are considered "**wait time**", where work is sitting in a queue.

NEXT [3]	DEVELOPMENT [4]		FUNCTIONAL TESTING [3]		WAIT FOR RELEASE IN TEST ENV. [3]	RELEASE TESTING [5]	WAIT FOR RELEASE TO PROD [∞]	DONE
	doing	done	doing	done				

VALUE (touch time)   QUEUE (wait time)

Flow Efficiency on the board: only some states are value-adding

Knowing this, calculating flow efficiency for a story is straightforward: $flowEfficiency = \frac{touchTime}{touchTime+waitTime}$ (Lowe2015). An easy way to think about it is: over the total time that the story has taken from start to end, what percentage of that time was spent in states where the story was actually being worked on?

To calculate flow efficiency you need to know how long stories are spending in each state of your process. The easiest way to measure this is to record transitions between states: record the date when a story moved from one state to another, and then you can easily calculate how long it spent in each state.

Once we know the flow efficiency for all stories, we can visualize it and calculate its average to find the flow efficiency of our process.

A high flow efficiency means that work flows through our process as fast as it possibly can, with minimal wait time or delay. It's the essence of agility: going from idea to delivery really quickly, to get feedback and adapt to change.

Flow Efficiency: how much potential improvement is there in our process?

Why is flow efficiency so low?

The picture above shows real data from Mattia's team. When they first saw it, they were shocked to see that their flow efficiency was a mere 29% - they thought they were a mature agile team! They spent a long time reflecting on their process and optimizing it, how could it be that they were still wasting 70% of the time on a story?

Chances are that if you calculate your own flow efficiency you'll find similar results - or possibly even smaller - as it turns out that low flow efficiency is very common. At a team level, it's often as low as 15% (Wester2016). When you expand this analysis to the whole organization and look at the total time for a piece of work from idea to delivery to customer - including all those things like project inception, business case, approval from finance, etc. - then flow efficiency is easily below 5%. A flow efficiency of 1% to 5% is common (Forss2013), and David Anderson says that "flow efficiencies of as low as 2% have been commonly reported by managers and consultants" (Anderson2013).

The reason for such low values is that most teams are still structured to optimize for *resource efficiency*, rather than for *flow efficiency* (Rothman2015). They focus on making sure that people are always busy, where idleness is seen as something to get rid of, something that can be made more efficient. People are often given specialized roles, because it's deemed more economically beneficial to have experts that focus on specific areas, in order to drive down the cost for the activity (e.g. tester, DBA, sysadmin). Work piles up while waiting for one of these experts to be available, only to then be passed on to the next required expert, which will lead to more waiting. Each step of the process is optimized locally, without noticing the effect that this has on the rest of the company.

When teams optimize for flow efficiency instead, the focus is on the work rather than the "busyness" of the workers. It's more important to complete work quickly than to keep people busy. These teams often use WIP limits to make sure they finish something before starting something new. There is some slack built into the system so that someone is always free to pick up a urgent unplanned piece of work and queues are avoided. People are less specialized and often help each other across different roles. These teams optimize their end-to-end flow, sometimes at the expense of making some steps suboptimal.

Managers often fall for optimizing for resource efficiency, thinking that it's more cost effective, but when you take a systemic approach and look at the end-to-end processes it turns out that flow efficiency is usually cheaper: overall lead time goes down (meaning that work takes less time to go from being an idea to being delivered); the team becomes more predictable (so you can make better promises and build trust with stakeholders); and you get feedback earlier (which you can act on to improve customer satisfaction).

Real-life example: here's how much we can improve

Our low flow efficiency shows just how much opportunity for improvements we have! Indeed, in lean wait time is called "waste", but you can also see it as "potential for improvement".

In "Out of the crisis" (Deming1982) Edward Deming famously noted that, in most processes, 94% of possibilities for improvements belong to the system and only 6% to the workers. Flow efficiency is a great confirmation of that finding: teams can't improve just by telling people to work harder - people are already doing the best they can, but even if they did improve they would only impact the "touch time", which is a small part. If you really want to improve you need to change your process to focus on flow efficiency.

If you look at the flow efficiency graph above, you might notice that one of the work items (towards the right side) is completely green. That was a due to a problem in production, whose resolution was expedited to take priority over everything else.

All hands were on deck to fix the issue as quickly as possible: some developers immediately started implementing a fix; a tester paired with another dev to write a script that would prove that the fix worked; the business analyst liaised with the ops team to line up the necessary people for a release to production. Some people were idle at times, waiting until they'd be able to contribute. Flow efficiency on that story was 100%, the work was never in a wait state. Now it was down to us to ask: what would it take to work like that all the time? What's stopping us?

Tips for improving flow efficiency

Here are some tips for improving flow efficiency in your team.

Visualize wait states. Make sure that wait states are represented on your board, so that it's clear when a story is being worked on and when it's waiting. For example, visualize columns like "dev done", "waiting for testing", "ready for release", etc. Once you visualize queues you can start tracking the time spent in there. You can start noticing when they're growing too much and do something about it.

Use WIP limits to make sure wait states don't grow too much. WIP limits will introduce stress to your system, making impediments to flow become more apparent. When your WIP limits are set low enough you're building slack into your process, encouraging collaboration whilst also ensuring that someone will be free if an emergency were to happen. Multitasking drops, resulting in higher quality. Team members get used to asking "Does anyone need help finishing anything?" before they pick up something new.

Reduce batch size. Take activities like planning or releasing and try to do them more often but in smaller quantities. For example: plan every week instead of every two weeks; invest in continuous delivery to release fewer stories more frequently; work on features instead of projects. Striving to do these things more often will highlight the cost of those activities (the so-called "transaction cost" Reinertsen2009), which will give you a clear indication of where you need to improve.

Encourage people to be cross-functional to reduce dependencies on any one individual team member (or on a dedicated functional team, like a centralized QA team). When people are cross-functional, anyone can pick up any piece of work, drastically reducing wait time for an activity.

Invest in quality to reduce the amount of unplanned rework. Invest in engineering practices like testing, pairing, refactoring, clean code, and any others that improve quality.

Time in state: "Where should we focus our improvements?"

When looking for ways to improve as a team, it's often hard to decide what areas we should focus on. What's going to have the most impact? Very often this decision is based on anecdotes, gut feelings, or pure guesswork. Time and time again teams choose to focus on reducing their touch time, for example they aim to "build faster" or "test faster", but don't realize that, even if successful, those improvements will only have a small impact.

The **time in state** metric is a good way to look for areas where we should focus our improvement efforts. For each story, it shows how long it has spent in each state of our process, as a percentage of the overall lead time.

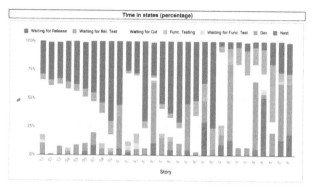

Time in state: where do stories spend more time? Where do we need to improve?

Do you see how the blue part (development) is usually very small? And do you see how the most prominent colors are yellow (waiting for cut), purple (waiting for a release) and red (next, meaning it's waiting for a developer)? Often this metric shows that development time is just a fraction of the overall lead time of a story. Stories tend to spend most of their times in wait states, and that's where we should focus our attention.

Reducing development time is usually very hard. It involves training, investing in modern engineering techniques, possibly adding more developers (but that tends to slow us down initially, and only shows benefits in the long term). Even if we did all of that and managed to reduce the development time, we would still only have an impact on the small blue part! We should instead tackle the other big areas, by changing our process to reduce wait time.

So next time your project is late and your manager suggests adding more developers, you can say "No thanks, that's not going to help. But you can let us release more often, that would help a lot".

This is, by the way, the perfect demonstration of what Deming was saying when he wrote that "94% of possibilities for improvements belong to the system and only 6% to the workers": it's by changing our process that we'll unlock big noticeable improvements, not by asking people to work harder.

Release time: "How long does it take for something to be released?"

"Do we release often enough?"

A common impediment to flow in many teams is **infrequent releases**. Stories pile up in a sort-of-done state, waiting for the next available release. This is usually because of a combination of:

- **Releases are deemed expensive**; they have a high "transaction cost". For example, they might require downtime, involve a lot of bureaucracy, need special people on call, and so on. This leads to the erroneous belief that it's more cost-effective to batch together multiple stories before doing a release, to make it worth the effort.
- **Releases are pre-scheduled** on a regular cadence, for example at the end of every sprint, every other sprint, every quarter, etc.

Although the cost of a release can be undeniable for some teams, we often forget to look at the other side of the coin: what is the cost of not releasing this story now? What's the cost of waiting for the next release? This cost is often referred to as "cost of delay" (BlackSwanFarming).

A good metric to highlight how much this delay might be affecting you is what we call **release time**:

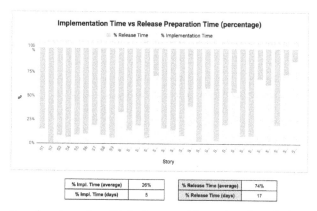

% Impl. Time (average)	26%		% Release Time (average)	74%
% Impl. Time (days)	5		% Release Time (days)	17

Release time: how long does it take for something to be released after it's been implemented?

For each story, we show in green the percentage of time that the story has taken to be "implemented" (go through development and testing), and in red the release time, which is the percentage of time that the story has taken to actually be released after implementation. The release time typically includes many administrative activities that contribute to making releases expensive, for example selecting the right version of artifacts, preparing release notes, synchronizing with other teams, creating change requests and getting them approved. Below the chart we also show the average release time, as it's useful to get a feel for just how much time is usually wasted before stories are released.

Big, infrequent releases have many disadvantages, and the more we bundle in a single release the more they negatively impact us. To mention just a few:

- The more change we releases at once, the higher the risk that one (or more) will fail.
- It's harder to diagnose which change has gone wrong.
- It takes longer for stakeholders to see the changes, so feedback is slower.
- Features that could help the company make money are not available to customers as early as possible.

- When problems arise, it's much harder to switch back to the context of a feature that we've worked on months ago.

We recommend the "Continuous Delivery" book (HumbleFarley2010) for more details and for ways to fight these problems, and the excellent "Accelerate" book (ForsgrenHumbleKim2018) for how these practices strongly correlate with high performance in teams.

With this metric, you have data to show how long stories usually wait for. You can use it to have a conversation about the effect of the decision to release so infrequently, and hopefully to drive changes towards continuous delivery.

Real-life example: from two months to a few days

In a team that Mattia was working with, releasing code took nearly two months. They were working on an old codebase which, for various technical reasons, required downtime to be released, and the company couldn't afford downtime during the day as that would mean product sales would stop.

Releases had to be scheduled during the night, with representatives from various teams on call (dev teams, ops, DBAs, etc.). All these people would have the next day off in lieu. Because of this big overnight process, releases were considered very expensive and were only scheduled once a month, with no appetite to do them more often. The topic had become a bit of a taboo around the office, and there was a general feeling that "things are this way and they can't be changed".

Mattia and his team generated the following "release time" chart, which showed that nearly 90% of the lead time of stories was spent waiting for the next scheduled release:

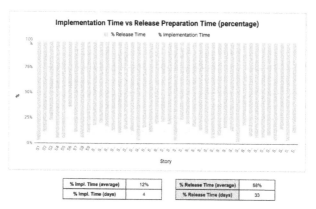

% Impl. Time (average)	12%		% Release Time (average)	88%
% Impl. Time (days)	4		% Release Time (days)	33

Real-life release time: nearly 90% time is spent waiting for release

After printing a big A3 version of the chart, they went to have a chat with the release management team. Pointing to the data, they explained how this long delay was affecting the dev teams, and ultimately the company. Seeing the data, the release managers started to understand the scale of the problem. Nevertheless, there wasn't an easy way to reduce the release cost with the current state of the codebase. It became clear that they couldn't just do the same style of releases more often, but instead they needed to invest into continuous delivery practices.

It was a difficult conversation, but having data at hand shifted it from the usual opinion-based confrontation with no results to a data-driven discussion that actually helped things move forward. With the blessing and support from release management, when the next big project came in, the development teams invested in a modern infrastructure platform using Docker and Kubernetes, to allow them to do daytime releases without downtime. Together they set up a new process so that releases could be scheduled on demand, as quickly as the next day.

4.6 Public resources

Troy Magennis has a collection of free spreadsheets to cover many metrics needs. Some about flow metrics in particular are "Flow Metrics Calculator" and "Team Dashboard".

Find them at http://bit.ly/SimResources

The "Penny Game" (Skarin2008) is a simple but very effective exercise that you can do with your team to see firsthand the effect that different batch sizes have on the flow of work. We've used this quick game countless times and it always generates great "aha" moments in the team.

4.7 Get started!

- **Start recording transitions between states**. Note down the date when a story moves from one state of your process to another.
- **Generate a CFD** for the last few months (a shorter period of time is fine if you don't have enough data). Which patterns described in this chapter do you recognize in your CFD? Think about the suggested actions in this chapter or others than can help improve your situation.
- **Plot your Net Flow**: is your process balanced? If it isn't, consider introducing WIP limits to force finishing work in progress before starting new work.
- **Calculate your flow efficiency**. Discuss with your team your current efficiency, and how you could use some of this chapter's suggestions to improve it.
- **Calculate your release time and time in state**. Reflect on whether you're releasing often enough, and where you should focus your improvement efforts.

4.8 Summary

Flow is the movement and delivery of customer value through a process. Having good flow means that we deliver value at a steady, regular pace. The team has predictability and the amount of work in progress is balanced to the capacity of the system.

We can use metrics like CFD and Net Flow to observe our flow and spot impediments to it, giving us a chance to improve. Flow efficiency tells us that we can make our flow better by focusing on reducing wait time and working in smaller batches.

Looking at our release time we can evaluate if we're releasing often enough. We can then analyze the time in state to decide where we should focus our attention.

5. Metrics for Quality: How do we know this works?

> ## Key points
>
> - Quality is 'conformance to requirements'
> - Use the RAS model to look at dimensions of conformance
> - Find people affected when conformance is not met
> - Agree metrics and tolerances with these people that measure conformance
> - Use these metrics to help prevent failure demand
> - Ensure the quality of your metrics practice
> - Pay attention to costs
> - Use your metrics to tell compelling stories

This chapter focuses on **quality**, or "conformance to requirements".

5.1 Why should I care about metrics for quality?

As a **team member**: by using metrics for quality you get a better idea of whether what you are spending your time doing is helping to improve the product and people's experience of it.

As a **Scrum master** or **coach**: you have data points that prompt useful conversations about team practices and behaviors, and that you can use for running data-driven retrospectives.

As a **product owner**: metrics for quality give you different lenses for looking at what makes your product fit for purpose.

As a **manager**: metrics for quality give you an indication of whether the team's effort is having the desired effect.

As a **customer**: metrics for quality allow your requirements to be expressed in a way that better matches your needs.

5.2 What do we mean by quality?

In this chapter quality means "conformance to requirements", a definition taken from Philip B. Crosby's book "Quality is Free" (Crosby1987). In that book, Crosby says:

> When all criteria are defined and explained, then the measurement of quality of life is possible and practical. In business the same is true. Requirements must be clearly stated so that they cannot be misunderstood. Measurements are then taken to determine conformance to those requirements. The nonconformance detected is the absence of quality. Quality problems become non conformance problems and quality becomes definable.

So, how do we clearly state those requirements in a way that cannot be misunderstood? And then how do we take measurements to see if what we are building conforms to those requirements?

One of the challenges of working with software is that it is invisible. Don Reinertsen in "The Principles of Product Development

Flow" (#Reinertsen2009) quotes an engineering manager at Hewlett Packard (HP):

> Our inventory is bits on a disk drive, and we have very big disk drives here at HP!

So far in this book we've looked at two dimensions that make our work visible. We've looked at how long it takes for the team to get the work done:

- Throughput - the rate at which the team completes work.
- Lead Time - time taken to complete a given piece of work.

This starts to make what is unseen seen. It also gives us our first visibility of quality, of conformance to requirements. It tells us if we are meeting the requirement to deliver software in predictable time scales. Is the software going to turn up when we said it was? And is it going to arrive in time to be of any use.

That's a start. In this chapter we'll identify a number of different dimensions we can quantify and measure. We'll see how the metrics we get by doing this help us make decisions about how the team works, what gets built, when it gets built and how it behaves when it is operational.

Our starting point for this is the IBM RAS Model (RAS). RAS Stands for Reliability, Availability, Serviceability. Thinking in terms of these 'itys', and there are many more that can be added, gets us thinking specifically about what different people's needs are. 'How reliable do our customers need our service to be?', 'When do our customers need our service to be available', 'How long does it take for our support team to get the service back up and running when something goes wrong?'

5.3 Use quality metrics to identify the needs of various groups of people

We find Crosby's definition of quality as 'conformance to requirements' helpful because it makes an often slippery subject tractable. It gives us two questions to ask; 'What are the requirements?' and 'How do we know if we are meeting them?'

The functional or behavioral requirements of the software the team builds will depend on your context. We work closely with our Product Manager and use methods such as Specification by Example (#Adzic2011) to ensure we are meeting these. Having an executable specification, with tests that are run as part of the build cycle gives us a pair of useful quality metrics - number of scenarios tested, number of tests passing.

We go into metrics for function and behavior in the chapter on value 'Why are we making this software?' Here we want to look beyond that. Conformance to requirements is not just about function or behavior. To identify and explore it's other dimensions we can borrow a model from IBM. It originates in Mainframe hardware but systems thinking has since extended its use to services in general. The IBM RAS Model gives us three lenses for looking at other dimensions of conformance to requirements:

- Reliability - Does your software consistently do what people expect it to do?
- Availability - Is your software available at the times and in the places where people need it?
- Serviceability - When things go wrong with your software, what does it take to get them fixed?

Why is this model helpful? It encourages you to ask questions about people's quality expectations and how you will meet them.

It encourages you to ask why. Why do you need this to be reliable? What is an example of unreliable behavior? When does it need to be available? To whom? How quickly? When it goes wrong how do you know it is broken? How quickly can you get it back up and running again?

These are all questions which can be answered in terms of metrics, metrics which can be tracked to tell whether the quality of the system, and the service provided, is good enough, getting better or getting worse.

Here's an example of the model in use. This is a diagram Chris shared with his executive team to show how a focus on functional requirements had meant the software a team had built was lacking in other dimensions:

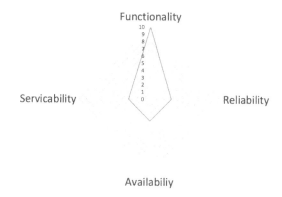

Functionality, Reliability, Availability, Serviceability

This was used to get executive approval for investing in leveling up RAS rather than adding more functionality.

We'll now look at these three RAS dimensions, and how you can measure them, in more detail.

Reliability

The first dimension of RAS is **reliability**, and reliability is all about trust:

- Can I trust that your software will do what you say it will?
- Can I trust that it will do this consistently with my expectations?
- Can I trust that it will do this predictably when things change?
- Can I trust you with my data, my privacy, my livelihood?

One of the reasons we favor forecasts over estimates is because we find them more reliable. We trust them more. Starting with building reliability into our operational metrics we build reliability into what we deliver. Does it do what the customer needs it to do? Does it do it consistently? These questions drive us to look at what we can measure to give us evidence that allows us to answer them.

We can use metrics to answer questions about reliability in three ways: consistency of service, predictability of service, and privacy & security.

Consistency of service

Software is written to meet a business need. To be of value it needs to do this consistently. Imagine a calculator which gave a different response every time it was asked for the answer to the question 'What is 2 + 2?'. It wouldn't get much use.

To ensure consistency of service we need to specify and test our software to work using representative data and context. We then need to identify meaningful indicators that it is behaving consistently in response to them.

Take the example of an online video service offering both five-minute cartoons and two hour movies. A customer of this service expects both the cartoon and the movie to start within a few seconds of them pressing play.

How do we know this is the case?

First we can look at the quality of the tests. How many test cases have been created for this scenario? If it is just one, play a 30 second test clip (Big Buck Bunny) then we might have a problem. A 30 second clip is great for testers and developers but is it representative of our customers usage needs? What happens with a 5 minute video? A two hour one? What happens if we change the video resolution from Standard to High Definition? What if the customer uses our service on both their phone and their Smart TV?

Video	Duration	Resolution	Device
Big Buck Bunny clip	5mins	780p	Phone
Big Buck Bunny clip	5mins	1080p	Smart TV
Friends	30mins	1280p	Phone
Friends	30mins	1280p	Smart TV
The Long Goodbye	120mins	1280p	Phone
The Long Goodbye	120mins	1280p	Smart TV

With a number of test cases in place we can now look at what metrics we expect to remain consistent across each test case.

Video	Duration	Resolution	Device	Expected Start Time	Actual Start Time
Big Buck Bunny clip	5mins	780p	Phone	3 secs	2.7 secs
Big Buck Bunny clip	5mins	1080p	Smart TV	5 secs	5.1 secs
Friends	30mins	1280p	Phone	3 secs	2.8 secs
Friends	30mins	1280p	Smart TV	5 secs	4.9 secs
The Long Goodbye	120mins	1280p	Phone	3 secs	2.5 secs

Video	Duration	Resolution	Device	Expected Start Time	Actual Start Time
The Long Goodbye	120mins	1280p	Smart TV	5 secs	4.7 secs

These tests give us metrics we can use to validate that the software we build meets the thresholds for consistency of service. We can use the same metrics in production to tell if it is continuing to do so.

Predictability of service

Related to consistency is predictability. The service is consistent if it behaves in the same way as it has previously; in the example above, videos always start within a few seconds.

The behavior of other aspects of a system might depend on inputs or events we have not seen before. In this case we can't test for consistency of service as we are dealing with unknowns. We can however use metrics and historic data to predict how the system should behave.

Staying with the online video service example, in order for a video to be made available to customers online it needs to be transcoded from its source format to a streaming format.

Unlike start time, which we expect to be consistent across all videos, we expect the time taken to complete transcoding to be different for different videos. It is dependent on a number of variables including duration, source resolution, key frame density, number of output streams required and so on.

Because of all these variables we don't know exactly how long a transcode is going to take but my looking at the correlation between input data and time taken to transcode we can give a forecast for how long similar videos are going to take to transcode:

Video	Duration	Resolution	Device	Transcode Time	State
Big Buck Bunny clip	5mins	780p	Phone	2mins	Completed
Big Buck Bunny clip	5mins	1080p	Smart TV	3mins	Completed
Friends	30mins	1280p	Phone	15mins	Completed
Friends	30mins	1280p	Smart TV	20mins	Completed
The Long Goodbye	120mins	1280p	Phone	100mins	Completed
The Long Goodbye	120mins	1280p	Smart TV	300mins	In Progress

This can be communicated to the customer. By monitoring the time taken to transcode we can spot if anything abnormal is happening and intervene. Have we discovered a new variable that affects transcode time? Has something gone wrong in the transcode process?

Privacy and security

Reliability is the system meeting users' expectations. As we have seen this includes the expectation that it will behave consistently for seemingly similar scenarios and that its behavior is predictable for things that are yet to happen.

Another expectation is that the users' data, their interactions with the system and the system's behavior in response to both of these will be private and secure. Any system that stores or processes customer data needs to demonstrate that it that data, and user actions on that data, are both private and secure.

As with reliability in general this is about trust. Can you trust that the people accessing your system are who they say they are? Can the people using your system trust you to look after their data and their privacy?

If you work on systems in regulated environments, for example health care or banking, or you work with customers who have their own security policies (such as broadcasters) you will need to present evidence for the security of your system.

Metrics can help provide evidence that your system is both private and secure. When working with two of the largest broadcasters in Europe, Chris was required to provide dates and results for penetration tests conducted on his team's system.

As with consistency and predictability the number of test cases being exercised is a good indicator of this. Examples of these for the example above include:

- Customer can see the videos they have purchased.
- Customer cannot see videos they have not purchased.
- Customer cannot see other customer's purchases.
- Administrator can see all customer's purchases.
- Administrator cannot see customer's payment details.
- Failed login attempts are logged.

Logging events such as failed login attempts or requests for non publicly available resources will give you insights into what attack vectors exist on your system and whether they are being exploited. This data can be used to make the case for investing in increasing security.

Availability

Having established how to tell whether your system can be relied upon by the people that use it we now look at how we ensure it is available when and where they need it: the **availability** of the system.

Measure temporal availability ("when")

When does the software the team produces need to be available? The graph below shows the usage of a web-based video transcoding service over a 24 hour period:

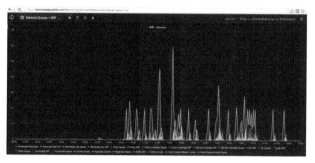

Measuring queue length to identify what availability you need

The spikes show when it was being used the most. You can see to the left of the graph that there is little or no usage. This was because it was night-time and no one needed the service. This data was used to plan the uptime of cloud compute resources, allowing them to be switched off when there was no demand.

Measure geographic availability ("where")

Where do people need to have access to the software the team produces? Today most software is distributed over the internet and so is theoretically available everywhere. In practice there is work to be done to make software available around the world. Localization to language and regulations is one part of this. Another is the infrastructure required to provide the same level of availability to all users. For example, a website served from a cloud provider's region in Europe will not be as responsive to users in Australia. There are of course technologies to address this: Content Distribution Networks (CDNs), for example. Unless we are measuring the quality of user experience in different geographies then we don't know how available our service is to these people. What is the threshold of latency for page loads or API responses that needs to be met for a service to be considered available?

Measure the effect of scale on availability ("how many")

Does the software the team produce remain reliable and predictable as more people use it, or it is put under higher workloads?

As use of your service increases, does it remain available to everyone or does performance degrade? Does reliability start to suffer?

Chris saw an example of this where and API used to ingest data from third party sources worked fine when the data payloads were in the low kilobytes but as they increased in size, rather than ingest requests being refused or performance degrading gracefully, the system ground to a halt and worse still, the impact of one user started to affect others, regardless of the size of the payloads they were submitting.

The solution was to put in place in-production load testing, run after every new release of the software. These demonstrated the operational tolerances under which the system could safely operate.

The next step was to put in place rate limiting to prevent users from overloading it. This is not the end of the story though. Where demand exceeded capacity this was an opportunity to increase capacity for users who needed it. This lead to work to implement autoscaling.

Serviceability

We've now established the parameters in which we can make assurances about the operational quality in terms of the consistency and predictability of the service you provide - its reliability. We've looked at when and where we can make this service available to people - its availability.

We've put measures with tolerances and thresholds in place for both these dimensions which together give us a good idea of what a working system looks like. What do we do when something happens to take us outside of these operational tolerances? Ideally we'd like our systems to just work and for our end users to have no need of assistance. What can we measure to tell us whether we have achieved this? If we have not, how can those measurements help us improve?

To answer this, let's look at measuring our **serviceability** (also known as supportability). Serviceability is about building your software with the expectation that things will go wrong; building with failure in mind. However well you may know your software, you don't know, and cannot know, everything about the world in which it will operate.

In 'The Logical Thinking Process' (Dettmer2007), William Dettmer models this as The Span of Influence, The Zone of Control and the External Environment. If something isn't under our direct control, what can you do to influence it? If or when something breaks, how can we get it back on track? To influence we need visibility; we need to know what is happening. We need indicators that something is not right so we can do something about it. If something is beyond our influence, in the external environment, how resilient are we to things going wrong there?

The metrics we put in place for reliability and availability go some way to providing visibility. By the same token, we see the impact of poor serviceability on our reliability and availability.

So how do we know what the quality - the conformance to requirements - of our Serviceability is? To answer this we need to dig into what is required to recover from a lack of availability, i.e. to get the system back up, or a lack of reliability, i.e. to get the system to do what the customer expected it to do.

Timeline of an outage

Below is the timeline of an outage, where a system became unavailable and due to a lack of serviceability took days to return to a state of reliable availability.

A sales demo had been built using preview technology from a cloud hosting provider. This demo was shown at a major trade show. During the trade show the demo stopped working. The ops team on duty were not able to recover the service themselves.

Date	Time	Event
May		Sales demo built using preview tech from cloud hosting provider
01-Sep		Cloud hosting provider advises of roll out of changes
04-Sep		Trade Show added to Demo Calendar
15-Sep (Saturday)	09:30	Sales Team report demo at trade show not working
	10:02	Support ticket raised
	10:04	Ops team start troubleshooting
	10:10	Ops team say input needed from outsourced dev team
	11:00	Sales phone VP Eng
		VP Eng gives sales outsourced team lead's phone number
		Sales phones outsourced team's lead
	12:26	Developer from outsourced team joins Slack conversation
	...	
	16:20	Support ticket opened with cloud hosting provider
	16:46	Developer from outsourced team discovers 11 Aug breaking changes announcement
	16:53	Message left with cloud hosting provider support agent
	16:55	Developer shares discovery on Slack
	...	
	19:14	Complete reload of production code
	20:00	Discover email response time for cloud host is business hours only on current SLA
	21:00	
	22:00	
	23:00	
16-Sep (Sunday)	00:00	
	01:00	
	02:00	
	02:47	Cloud host responds to support ticket ref breaking change announcement from 1 Sept
	...	
17-Sep (Monday)	17:00	Cloud hosts supply instructions for fixing the problem
	18:00	Upgrade to production
	19:00	Load Testing
	22:30	Outsourced team reports everything is back to normal

Timeline of an outage

To identify the source of the problem an outsourced development team had to be contacted who in turn had to contact a cloud hosting provider. This meant that although it only took an hour to fix the problem, it took a total of 61 hours to recover the service:

Time to recover service	61 hours
Time to identify problem	55.5 hours
Time to fix problem	1 hour
Time to test fix	4.5 hours

It's the top line figure, the time to recover the service or 'Mean Time to Recovery' when we are looking at the average across a number of incidents, that we are interested in. Beware of vanity metrics here, time to respond to a support call is of little or no use if the time to resolve the issue is an order of magnitude higher.

To bring an awareness of how serviceable your software is, cycle people from development and test through the support desk. Even better, have people go out and sit with users of your software in the wild.

A further metric of serviceability is the Number of manual 'How Tos' the support team require to deal with support requests. If you have this on a wiki you can keep a tally of it and use it as a source of requirements for improving the system. You can also see whether it is growing or shrinking over time!

Liquidity

In the serviceability example we've just looked at, the reason it took so long to get the system up and running again was the lack of people who knew how to fix the problem. This can be addressed by measuring the team's liquidity; how easy is it for people to move from one area of the codebase to another?

To start with, find out who can do what in your team. Do you have parts of your code base that only one person can work on? What do people *want* to work on? Are there people who are dedicated frontend or backend developers? What about operations? Are they part of your team or separate?

A tool that helps answer these questions is a Skills Matrix. It also gives you data you can use to inform decisions about how you organize around the work. That is, who does what.

The idea of a Skills Matrix comes from Chris Matts (Matts2013) Here's one from a team Chris worked with:

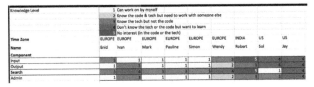

Knowledge Level		1 Can work on by myself								
		2 Know the code & tech but need to work with someone else								
		3 Know the tech but not the code								
		4 Don't know the tech or the code but want to learn								
		5 No interest (in the code or the tech)								
Time Zone		EUROPE	EUROPE	EUROPE	EUROPE	EUROPE	EUROPE	INDIA	US	US
Name		Enid	Ivan	Mark	Pauline	Simon	Wendy	Robert	Sol	Jay
Component										
Input		3	1	1	1		1	3	5	4
Output		1	3		1	1	1	2	3	4
Search		3	4	3	3	3	4		1	4
Admin		1	3	1	1	1	2	3	3	4

Skills Matrix

To create the matrix the team were asked 'What skills are needed to work on our software?' This give you the Y axis on the matrix. The team members make up the X axis.

Members of the team were then asked to rate themselves on how their skills matched up with what was needed using the following scale:

1. I can work on by myself.
2. I know the code and tech but need to work with someone else.
3. I know the tech but not the code.
4. I don't know the tech or the code but want to learn.
5. I have no interest (in the code or the tech).

This information was then used to identify where liquidity was poor, in the example above, the only person who could work on the 'Search' component was Sol, based in the USA. It was the search component that failed in the outage example above. An action from looking at the liquidity matrix was to do some pair programming between Sol and other team members.

5.4 Use quality metrics to discover who is affected by quality

The RAS model gets us thinking about quality in a number of dimensions. But why should we bother with this?

Hopefully we know why we meet functional requirements. Someone in the business, a product manager or a customer has a specific purpose, something they cannot do which they want or need to do and a technology solution has been selected. Surely this alone should keep the team busy? Making the system reliable, available and serviceable on top of this sounds like a lot of work.

To figure this out we need to know who is going to be affected when these requirements are not met; when they are not conformed to.

To some people quality is invisible, or rather it only becomes visible when it is lacking. Customers expect a product or service to 'just work', and why shouldn't they? So the stuff that requires effort and expertise to get right isn't counted. 'Why aren't you working on feature X?'

For others quality is all too visible: front-line sales staff demoing your product, people on the support desk, the customers themselves who have to use the damn thing; all their lives are, invisibly, impacted by quality.

We can find this out by including RAS in our requirements discovery.

Example: discover people affected by quality

Find people who have an interest in, that is are affected by, the features and functionality of the software the team produces. Ask them what they need in order for their needs to be met in terms of Reliability, Availability and Serviceability.

We've segmented the audience for stories of quality into three groups:

- Producers - The Team or Teams that make the product, system or service. This includes Product Managers, Developers, Testers etc.
- Operators - The people who keep the product, system or service running. This includes the people on the Help Desk, Tech Support, Site Reliability Engineers, Sys Admins
- Users - The people who make use of the product, system or service. This includes Public Customers as well as Internal Users such as Sales People, Finance etc.

In doing this don't forget yourself. How do you define the quality of your work? How do you define the quality of the practices and environment in which the work takes place?

RAS	Requirement	Required by	Who is a	Because...
Reliability	Videos start quickly	Home customers	User	I've paid for my movie!
Availability	Website is always available.	CEO	Producer	It's where we do business
Serviceability	I know as soon as something is offline	Support Desk Staff	Operator	Otherwise I can't help our customers

Now take the answers and then ask 'What evidence do we need to tell we are conforming to these requirements?' How good does what we build need to be in order to meet people's neeeds? By talking to the people affected we can identify metrics and set thresholds for when a change in that metric indicates non conformance to requirements. In doing this we should also identify a secondary metric to sanity check the primary one.

RAS	Requirement	Primary Metric	Secondary Metric
R	Videos start quickly	Time of start event minus time of play event < 5 secs	Internet connection is at minimum supported level
A	Website is always available.	0 downtime alerts/week.	Website traffic is at expected level
S	I know as soon as something is offline	Time I'm paged minus time of downtime alert < 1 min	

5.5 Use quality metrics to identify and reduce Failure Demand

By speaking with the people our software affects we know what they expect from it. By using metrics with tolerances and setting thresholds we quantify this. By testing against these thresholds and tolerances in development and raising alerts when they are breached in production we are able to identify and prevent 'Failure Demand': *failure demand is demand caused by a failure to do something or do something right for the customer'* (Seddon2008).

Identify Failure Demand

There are several ways to identify failure demand when writing and running software; these are covered in this section.

Log the abnormalities

How do we know if the system is doing what it supposed to? We need to make its behavior observable. We do this through logging and alerts. Alerts warn us that something has gone wrong or is about to go wrong; a server is under abnormal load for example or a queue of jobs has exceeded a safe limit.

Let's say we have some video processing software that uses queues to analyze videos. We want to know when the system is under load of different kinds. We can visualize the state of the queues as a Cumulative Flow Diagram (using a graphing tool like Grafana) to show the current queue states. For example, if the video ingest queue has grown to an unhealthy size, we might see a CFD like this:

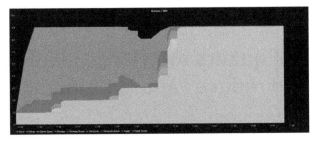

Ingest Queue (dark green)

These logs and alerts tell us what happened, what a user did that caused an unexpected outcome, but they don't tell us what the user expected to happen. For that we need to solicit user input, we need people to contact support and tell us what went wrong.

Track the defects raised

Tom DeMarco says: 'Any organization that fails to track and type defects is running at less than its optimal level'. This certainly holds for defects found in production; we'll come back to this when we talk about metrics for testing below. Defects found in production are a measure of failure demand. They represent a failure to conform to requirements and so a failure to do what people need our software and systems to do.

This categorization of defects is significant as it tells us about the modality of our work. Is it all the same or are there attributes common to groups of defects that can inform the way we work?

Decide whether to contain or fix

When responding to a defect there is often a choice between containing the problem to get the system running again or the customer problem addressed and fixing the underlying problem.

At Google, Site Reliability Engineers have an 'Error Budget' the use of which triggers the switch from contain to fix:

> "If the SRE team that looks after a particular application or service finds that it is spending more than 50% of its time doing manual operational work to address problems in the software, the development team must pick up the slack." (Skelton2018)

Prevent Failure Demand

By identifying failure demand in the wild we can fix our software and stop it failing in the same way again. But wouldn't it be better if that failure demand wasn't there in the first place? When we are responding to failure demand that is time and effort that is not going into providing value.

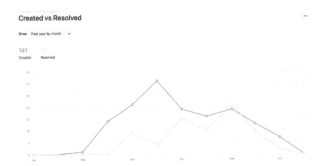

Graph of support tickets raised and resolved over the course of a year

The red line shows tickets raised, i.e. demand. The green line shows tickets fixed, i.e. throughput devoted to fixing defects. That is throughput that is not delivering value.

Instead we can make the business decision to invest in preventing failure demand in the first place. As John Seddon puts it, 'turning off the causes of failure demand is the most powerful economic lever available to a manager' (Seddon2008).

Capture RAS in requirements

When we discussed Reliability, Availability and Serviceability earlier we used examples of different people's requirements for RAS in order to specify the operational tolerances of the software the team builds. We recommend including these in specific examples used to discover and specify requirements. An effective 'Specification by Example practice' helps the team understand the different dimensions of quality and provides a foundation for building software that works within the operational tolerances established for them.

Write and run meaningful tests

The corollary of an effective Specification by Example process is automated tests that reflect the way in which you expect the system to be used.

When looking at your team's code here are some questions you might want to ask:

- When did the team do their first test with representative data?
- Has the system been tested under the loads expected of it?
- Has that load testing taken geographic availability into consideration?
- Do your tests take into account the different people who are affected by non-conformance to quality? For example, do you have tests to check you are writing meaningful information to logs for people on support to work with?

Choose metrics for testing carefully

Can we put any metrics around our testing practice? Crispin and Gregory urge caution here. They suggest that you look at each metric's return on investment and then decide whether to track or maintain it. "Does the effort spent collecting it justify the value it delivers" (CrispinGregory2009).

A good example of this is defect rates *during* development: "There's not much value in knowing the rate of bugs found and fixed during development, because finding and fixing them is an integral part of development" (CrispinGregory2009).

What is worth looking at is the number and quality of tests. We talked about this in the sections on Reliability and Availability above. Tests define and validate the operational tolerances of the system. An area of the system that is light on these is likely to have loosely-defined operational tolerances. Deciding how tightly these need to be defined will depend on your context.

5.6 Use quality metrics to tell compelling stories to the right people

Metrics do not grow on trees, waiting to be harvested. Metrics are made from data and this data is generated by tools, processes, and instrumentation. How do we know this data, and so the metrics we derive from it, can be trusted as the basis of our business decisions? We can look at it through a number of lenses to see if it is fit for this purpose:

- Provenance: Where does this data actually come from? Was it recorded by hand? If the data is collected manually is this done so accurately and consistently?
- Cost: What is the overhead of producing data used for metrics?
- Cleanliness: Is the data correctly categorized? Is there ambiguity in the data?
- Selection & Bias: Is there bias in the selection of data?
- Visualization: Is the visualization of data impeding people's understanding?

Even with clean data of known provenance, scruitinized for bias and selection, and presented in a clear visualization, you are not dealing with objective fact. You are telling a subjective story. If that story is going to help people make business decisions it needs to be interesting enough for them to listen to it. It needs to be memorable and it needs to be actionable. To make sure that the stories you tell achieve this, lets look at some of the characteristics and pitfalls of story telling.

"What exactly is a story? At a fundamental level, a story expresses how and why life changes" (Knaflic2015).

This succinct definition of what a story is, made by Cole Nussbaumer Knaflic, in her book 'Storytelling with data' is, like the metrics we talk about in this book, about the passage of time. All the metrics we have presented only make sense over time. They make sense because they show that things have changed. The data we choose to present is the 'how' life changes. The interpretation we attach to this is the why.

Knaflic goes on to say that we can use stories to 'engage our audience emotionally in a way that goes beyond what facts can do'

1. Set the scene - building the context for your audience. Why should I pay attention to this? In the example, this is the level of failure demand caused by the current code base. *This is an imbalance, why is this a problem? What are we going to do to fix it? What is our recommended solution?*
2. Having set the scene, you then go on to set out "what could be" if the solution proposed is accepted.
3. The story ends with a call to action. "Make it totally clear to your audience what you want them to do with the new understanding or knowledge that you've imparted to them."

Example: a metrics story with data

Here is a story that Chris told to an executive team to inform a business decision about what a team should work on. Following on from the outage discussed earlier in this chapter, the search component of the system the team were responsible for continued to cause problems. This component was unpopular with the team as it had been written by a third party and inherited by them. It was not written in the same language as the rest of the system and the consensus in the team was that it should be re-written by them. This view was contrary to that of the executive team who saw this as needless duplication of effort. Their preference was to add more features and deal with defects as they came up.

The story Chris told covered the three dimensions of Reliability, Availability and Serviceability. He started by showing this graph:

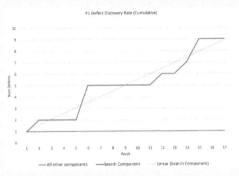

Cumulative Defects Discovered

To show that in contrast to the rest of the system, the search component continued to yield new errors. He then showed this graph of increasing search latency:

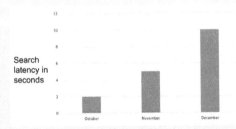

Search Latency

This shows that the system's availability was not scaling as more data was ingested into it over time. Finally he showed this line from the skills matrix we looked at earlier:

Search Serviceability

This shows how little knoweldge of the inherited codebase there was in the team and how this impacted its Serviceability.

Having presented the evidence to support the argument for re-writing he then showed the team's throughput graph:

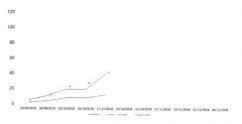

Throughput Forecast

This shows that the team delivered new features at a predictable rate and so the re-write of the search component could be forecast to be completed within a set amount of time. The clear articulation of a need, coupled with a commitment to complete the work by a set date meant that the executive agreed to the work being done, knowing that they could get back to adding new features after that point.

5.7 Debunking some myths

"Quality is for QA to worry about"

Whilst testing requires skills and abilities which may be the preserve of specific individuals in the team, quality is the concern of everyone. Because we define quality as 'conformance to requirements' we have to start thinking about it from the moment those requirements are known to us. Knowing what requirements we need to conform to if we are going to create a quality product or outcome we can then make sure everyone in the team, be they analyst, designer, developer, tester, operator, technical author or whatever other role you have, is involved in discovering what those requirements are, in all their multi dimensional glory, and ensure that the collective team effort is directed towards meeting them.

"We can add the metrics collection later"

You will almost certainly never get round to adding metrics later. If you are focused on adding functionality, and are not thinking about reliability, availability or serviceability, then you are not going to make the time to add the instrumentation to surface the metrics needed to measure these aspects of quality. Plus it will be a lot harder to add them in after the fact. Make the case for spending time and money on doing it in tandem with adding features. That way you will be able to tell whether the features you add are doing so without impacting the other concerns that ensure your product is fit for purpose.

"The customer is king"

Allowing the customer to drive all decisions is bad. First of all this assumes that the customer is a 'he'. This is just the start of how this myth excludes people affected by the product the team produces from an understanding of quality, of conformance to requirements. The customer may be the end user why pays to use your product, or it may be someone else, an advertiser for example who enables your product to be free at the point of use. But requirements and so quality do not end there. Any one whose behavior is affected for better or worse should be considered when you are looking at quality. This includes operations and support, marketing and anyone else within your span of control or sphere of influence.

5.8 Get started!

- As a team, look at the functionality of the software you produce. Discuss what the requirements are for its Reliability, Availability and Serviceability.
- Identify the people whose needs inform the RAS criteria. Discuss what the impact is when Reliability, Availability and Serviceability are not 'good enough'.
- Agree metrics and tolerances with these people that you can use to measure conformance.
- Look at the level of failure demand the team produces. How do you know when failure demand is being generated? Discuss ways you can help prevent failure demand in the first place.
- Find out the costs, both financial and otherwise, of running the software your team produces.
- Find one thing you as a team want to change, together come up with a compelling story that you can tell to help bring about that change.

5.9 Summary

In this chapter we've identified quality as conformance to requirements. Conformance is not a one dimensional 'does this work or not?' property but multi-dimensional. Those dimensions are identified when we discover how not meeting requirements affects different groups of people.

Working with these people we agree metrics which we can use to measure conformance to requirements. Knowing the conformance to requirements we then identify the things that are preventing us from achieving this: 'failure demand'. If we can eliminate the causes of failure demand then we can use all our capabilities to meet people's needs.

Of course, none of this happens in a vacuum. In any organisation (for-profit or non-profit), costs and budget will be a constraint on what you can or cannot do. In addition to financial cost, you need to be mindful of the cost of your work on the team itself and the quality of their working practices.

The quality of your measurement practice itself will affect how well you can do all of this. The outcome of having the means to measure the quality of your work and ways of working is that the team can tell stories, to themselves, the wider business and customers to better inform business decisions. These stories might celebrate success, warn of the impact of doing or not doing something, or alert people to an emerging trend.

Engage with all of this and you will be building a measurement practice that meets the requirements of the business.

6. Metrics for Value: Is this worth doing?

Key points

- Make sure the team can answer the question 'Why are we doing this'?
- Lead by example - have your operational and quality metrics in place first.
- Create directly observable usage metrics.
- Use Pirate Metrics: Acquistion, Activation, Retention, Referral and Revenue
- Look at how the needs of multiple customers and your capabilities interact.
- Expect to keep learning and adapting.

This chapter focuses on **value**, the worth of the output and outcomes for the user or stakeholder.

6.1 Why should I care about value metrics?

As a **team member:** by using metrics for value you can see the effect of the effort you are expending and take an active role in working on stuff that matters.

As a **Scrum master** or **coach**: metrics for value help you and the team focus on the things that matter.

As a **product owner**: metrics for value give you the means to tell whether your product is worth the effort expended on it.

As a **manager**: metrics for value help you ensure the team is focusing its efforts on the stuff that is going to give the greatest benefit to customers, the business and the team itself.

As a **customer**: metrics for value help direct the product to a better outcome and benefit for you.

6.2 What are metrics for value?

Metrics for value help us qualify why we do what we choose to do. They do this by quantifying our work with measurable outcomes. Rather than pointing to the lines of code we have written or the new page on our website that has gone live we can point to the impact we have had, the change we have made that demonstrates the value we have created.

How does a team reach this metrics nirvana? One of the joys and problems with writing software is that it can be a lot of fun. It's creative. It's about "making stuff". So when someone comes to your team and says 'Can you make me an X?' - where X might be a new button on a website or adding AI to your product - it can be easy to dive in and start building.

Diving in immediately is really starting with "what": what someone thinks they want. If we do this we are in danger of being very busy without having much to show for it. Sure, we will have built the thing we were asked to build but how do we know whether it fulfilled any purpose? With so little time and no shortage of people with great ideas for how they think we should be using that time, we need to take care before committing to doing work.

You or your customer might have any number of reasons why they want to build or change a piece of software. Adding a button to a website sounds like a great idea but unless you have a way of telling how the world will be different, and one would hope better, as a result of this mighty button being available you will never know whether it was worth spending the time and money to add it.

Starting with "what" is the path of least resistance, but there is a better way: starting with "why". In this chapter we look at how we can find out the purpose and intent behind the work our teams are asked to do, and how we can measure the value of that work. We have two motives here, both key to greater value: not wasting time and doing something worthwhile.

6.3 Provide a foundation of operational and quality metrics

Despite the importance of "Why", we have come to "Why" at the end of the book, having first explored operational and quality metrics. Why is this?

> Nobody should care about measuring something if it doesn't inform a significant bet of some kind. (Hubbard2014)

As developers we don't want to be spending time on things that no one really wants or that end up never being used.

We as software developers or engineers need to convince our customers of the value of measuring the outcome of the work we do with them - its efficacy in achieving the purpose for which it was set in motion. We will have far more success in doing this convincing if we can point to an example.

The metrics we have covered so far in this book provide that example. Measuring throughput reduces uncertainty on how much work the team is likely to get done each week. Measuring lead time reduces uncertainty about how long different types of work are going to take. Measuring the rate at which defects are discovered helps tell us if the methods we use to write software are having the desired effect of producing working code rather than a pile of bugs.

We as a team can point to this, share it with our customers and show them how in our day-to-day practice, we use metrics to inform the decisions we make and so inform uncertain decisions. This is all good but using **only** quality and operational metrics can leave us in a comfortable but potentially dangerous or costly situation. Gojko Adzic sums this up nicely with a driving metaphor (Adzic2013):

> Imagine you're in a car, driving on the motorway. If your velocity is below 20-30 MPH, you know that something is wrong. There might be lots of causes for that, and you don't even have to know the root cause to decide on corrective action - maybe phone ahead to tell someone you'll be late, or look for an alternative route. Using velocity for that kind of process control is great.

Having operational and quality metrics in place means we can determine that the car is roadworthy, that it can get down the drive and on its way. But this is not enough:

> If, on the other hand, the car is moving at 60-70 MPH, that piece of information stops being so relevant. There are much more important questions: Are you on the right route? Do you have enough petrol to reach the destination? At that point, improving velocity at the expense of other things would be unwise, as you might fail to make the right turn or be flexible when the road ahead suddenly gets blocked.

So, how do we make sure we are headed in the right direction? We look for things that will change as a result of our work.

6.4 Measure usage: is anyone using it?

The team has specified, developed, tested and deployed a new feature. Apart from the new commits to version control, the successful build in the build log and the sighs of relief all round, how can you tell if doing this has made any difference to the world? The first thing you can look at, as it is within your span of control, is whether anyone is using this new feature.

If you make measurement of feature usage part of the feature itself, that is you don't put it live until you can measure its usage, then you have your first point of feedback on the feature's value.

If no-one is using the feature the team has built then this should be a prompt for a conversation about the value of that feature and why it was decided to spend time working on it.

Examples: TV commercials

Here are some real-life examples from when Chris worked with a cross organization Dev/Ops team to build a website for an online market (Young2014). The website was to correspond with a new TV commercial, and allowed shop owners in the online market to make a personalized version of the TV commercial including in it the name of their shop.

The videos created by shop owners were stored in Amazon S3. The team were able to use the metadata created when a new video was uploaded to S3 to create this graph showing how many videos had been created:

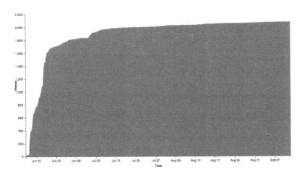

Videos Created by Customers

There was a Call to Action in the form of a mailout from the market website to all its shop owners telling them about the campaign and pointing them to the website where they could make their personalized ad. The initial growth coincided with this. A second email was sent out which coincides with the second bump in the graph. After that video creation tails off.

The second example is different. A new credit card billing feature for an existing SaaS platform consisted of 24 User Stories. The end-to-end time to deliver these 24 stories was 43 weeks, so roughly a story every two weeks. This was not a linear progression; as you can see from the graph, throughput displays the classic S-curve observed on many software projects:

—User Stories   —Credit Card Users

Throughput vs Usage

The graph also shows uptake of the new billing feature. A single user, who was in truth a member of staff for the company using his credit card to check the production system. That's a lot of work for a single user. This looks like the very definition of being stuck in the 'Build Trap' as Melissa Perri puts it (Perri2018).

Why was this? The business model shifted to working with OEMs rather than selling direct to developers. The graph below shows the number of OEM accounts created:

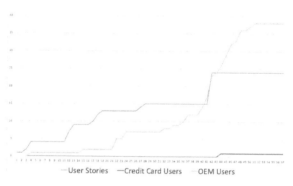

—User Stories —Credit Card Users — OEM Users

Throughput vs Usage

So all that time features were being added that were of no use to the system's users.

We've shifted the focus from what we're producing to its use but we are still some way away from answering the question 'Why are we making this software?' We're not making it in order for people to use it. It does not have intrinsic value. There is not a lot of value in measuring logins to a system if the person who has logged in doesn't do anything after that.

People use your software because there is some greater need or purpose. If we're not addressing this then use of our product will be short-lived. They will leave and go elsewhere. In the words of Peter Drucker:

> "The purpose of a business is to create and keep a customer" (Drucker2012)

What metrics can we use to measure how well we are doing this? More importantly, what metrics can inform decisions that will help us achieve this?

6.5 Use the Pirate Metrics: AARRR

To answer these questions, just like we used the RAS model for Quality, we'll use the excellent AARRR model from Dave McClure (McClure2007) to explore the creation and keeping of customers: Acquisition, Activation, Retention, Referral, Revenue (so-named because pirates go "AARRR"!).

Let's take a look at these one by one.

Acquisition

There isn't a lot of point building or adding to your product if people either don't know it's there or can't find it. It's a bit like lavishing money on the interior of a beautiful new shop that is located down a dark ally in an unloved part of town.

To get useful acquisition metrics you need to start with an idea of how many people could potentially be acquired as customers.

Where are you going to find these customers? How successful are different efforts to acquire customers? All of this may seem like it is outside the remit of your team. "Isn't this the job of marketing?", you might ask. Aren't they the ones who tell would-be customers all about the wonderful things to expect from your software?

If the marketing people are not part of your team, make them part of your extended family; bring them into your sphere of influence. Ask them for the metrics they use for targeting and acquiring customers. You can then reconcile these against the numbers who actually end up using your product and use this information to better inform how to acquire more of them.

It might be that there is no need to add new features to your product, rather you just need to be looking in different places to acquire customers.

To quote Peter Drucker again:

> "Because it is its purpose to create a customer, any business enterprise has two - and only these two - basic functions: marketing and innovation"

In the TV Video example we looked at above, usage increased in response to an email to the owners of the market's online stores directing them to the website where they could make their videos. Usage was, not surprisingly, directly correlated to these mailouts.

In the credit card example there was no marketing of the new feature and, not surprisingly, no usage.

Activation

By working with your new friends in marketing, people are now aware of your product or service. They have found you and now want to use what you offer to meet the need you address. How many of the people who make the journey to your website or download your app go on to use it? How many sign up, make a transaction or spend time using the resources you provide?

By tracking visitors to the site against sign ups you can get an indication of how effective your activation mechanism is and decide whether it needs more investment.

You can also look at how many of the calls to your help desk are to do with problems logging in or signing up. Is there a source of failure demand here?

Retention

People have found you and are using your product. Do they carry on using it or do they vote with their feet and go elsewhere? Either way, to make use of this information to inform business decisions you'll need to find out why.

For this you'll need to dig into the value proposition on which your product, service or business is based. This will be domain-specific but deserves to be understood. You can then ask people either why they have stayed with you or why they have left.

Go back to the business case that unlocked the funds or raised the money to make this thing happen. People will remain customers if they continue to get value from your product or service. We need to dig a bit deeper to figure out what that is.

Create a model of value

Let's use the example of Honeycomb here. Honeycomb, now part of Peach, works in video advertising. Once a TV commercial has been made, Honeycomb transcodes, packages and delivers the commercial from the Post-Production companies where the broadcast master copies are made, to the myriad TV station and on line platforms that broadcast or stream commercials to viewers.

Honeycomb's value proposition to customers is four-fold:

- Value for money - Honeycomb is cheaper than the competition.
- Ease of use - Honeycomb is simple and easy to use.
- Reliability - Post Houses and Broadcasters can see and trust the process.
- Range of destinations - Honeycomb can deliver to and comply with many destinations and their standards/regulations.

To inform decisions about how to prioritize work, the value of these four dimensions can be used. We can ask our customers which things matter to them, what things are missing. When we do this we not only get valuable data, we also get valuable marketing material.

Referral

We don't want the number of people using our product or service to go down, which is why we care about retention. We want the opposite, we want growth. An effective way of achieving this is referral. Here's retail entrepreneur Julian Richer on the subject:

> A word-of-mouth recommendation is always more convincing and is much better value than advertising. [However,] if people tell their friends about good service, they are even more likely to regale them with stories of bad service. Think of those 400 people hearing complaints about you for every one you hear directly... (Richer2017)

To test the role of referral in your acquisition of customers as part of activation, you can ask them how they found out about you. Honeycomb solicits feedback from its customers and, with their permission, includes it in their marketing:

Honeycomb Website - https://honeycomb.tv/

Revenue

Finally, let's talk about money. In order to continue creating and keeping customers, the way you do this needs to be sustainable. There needs to be more money coming in than going out. There needs to be a profit.

Profit is a function of us providing value to our customers. Julian Richer, the entrepreneur and founder of the UK's biggest hi-fi retailer puts the focus right back onto the customer:

> "The primary measure of a business' success should be customer satisfaction, not profits. Profits are simply an indicator that you are getting customer service right."
> (Richer2017)

So we need to measure revenue (a source of profit) and to be able to tell when our actions in pursuit of creating and keeping customers - the AARR of AARRR - are positively or negatively impacting this. Regardless of whether you see profit as a goal in its own right or as an indicator of your ability to satisfy your customer's goals you need to measure the impact of your actions on it. If you run out of money you don't have a business, let alone a profit.

6.6 Make your customers part of the decision-making process

When you have multiple customers using your team's software there is no guarantee that they will speak with one voice, nor that they will all want the same things at the same time. How do you decide which customers' needs to attend to?

If you are working with more than one customer, be they internal or external, make them or their representatives aware of your team's capacity constraints and make them part of the process for deciding where and when to focus on which of the company's shared value models.

Working with external customers

How do you decide which external customer's needs to prioritize over another's? One option is to use the 'friendliness' of the customer to determine priorities. How willing is the customer to collaborate with you on changes and decisions? The more collaborative a customer is, the more they are likely to be part of the learning and co-discovery process. They are probably more open to working with you on things for which other less amenable customers would set higher quality thresholds.

Working with multiple internal customers

The needs of external customers are often represented by a Product Owner or Product Manager, but does this make the external customer your only customer? You will almost certainly have internal customers too, such as Operations and Support teams, Marketing, Finance and your own team itself. If all of these internal customers have needs that are met by your team then there are business decisions to be made about whose needs get met, and when.

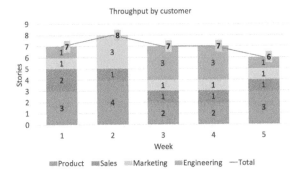

Division of throughput between work requested by different internal customers

One of the reasons for establishing operational metrics first is that you can use them to show your multiple internal customers that you are a capacity-constrained resource and that they need to work together to get the most effective use of your capacity. In the graph above we can see how a team's capacity to deliver work is split between four internal customers; Product, Sales, Marketing and the team themselves - Engineering.

Data allows you to demonstrate the fact that if you choose to do one thing you are choosing not to do something else. When it comes to pulling in new work to do, this team uses a queue with a maximum length of seven, the team's average weekly throughput, which the four internal customers can decide how to fill. The internal customers are able to make business decisions about which work to do next to benefit the company as a whole rather than only seeing their own concerns and so ending up with local optima.

You are your own customer

Don't forget that you and your team are customers of yourselves. You know when things need to be done to keep your software reliable, available and serviceable. Your 'Why' needs to be balanced against those of your other customers.

6.7 Expect to keep learning and adapting

Having value metrics in place tells us how we are performing against our model of what we think we should be doing. But what if that model is incorrect? Even if it was once correct, as the world turns around you it will become outdated as market conditions and people's needs change. As Tom DeMarco, author of the best-selling book *Peopleware*, says:

> 'Are we learning to do best what we shouldn't be doing at all?' (DeMarco1995)

This is why value metrics are so important. By treating your work as a series of experiments whose outcomes can be seen in changes to the value metrics, you can get feedback on how effective the team's work is. You and the people in your sphere of influence can decide whether to carry on with what you are doing or to change your focus.

By being mindful of the operational cost of features and their level of usage, you can also decide when it is time to remove things. Just because effort has been expended on building something doesn't mean it has to remain in production indefinitely.

6.8 Debunking some myths

"It's only the Product Manager who needs to care about the 'Why'"

Software engineering relies on a range of transferable skills that apply across different problem domains. There is an argument that if you are a software professional then the problem domain is a separate concern and not one that you need worry about. We argue otherwise. By understanding why software is being produced you will have a far greater understanding of the options available to meet that goal. In doing so you'll write software that gets used rather than merely deployed.

"Sales and Marketing are the enemy"

These post-it notes are from a retrospective Chris facilitated with a team:

Observations from a retrospective

They show a yawning gap between what is being sold or promoted to prospects/customers and what can actually be built. It is important to develop a strong working relationship with the sales and marketing people so that together you create a shared understanding of the problems to be solved and the opportunities to be realized.

"New features always add value to the product"

Programming is fun. It's a fulfilling, enjoyable activity. That doesn't mean that well written, tested and operationally ready code is actually of any use. Don't get too attached to your code. If a feature isn't providing anyone with value - as determined by your value metrics - then you should be happy to remove it.

6.9 Get started!

Here are some ways your team can get started with value metrics:

- Find metrics that show the three most used and three least used features of the software your team produces.
- Find the people in the business who care about these features.
- With these people, identify which of the Pirate Metrics (Acquisition, Activation, Retention, Referral or Recovery) these features apply to.
- For retention, determine what the underlying value model is that keeps people using your software.
- Together identify how you can measure the impact that these features have on the Pirate Metrics.
- Find out how much it costs to keep these features working. Look both at financial cost and at side effects.
- Talk about what you want to do with the least used features. Why are they not being used? Are they worth keeping?
- Talk about how what you know from the most used features could inform decisions about what features to add next.

6.10 Summary

To position the team for working with customers and the wider business on value metrics, get your operational and quality metrics working for you first. That way you have a working example to point to, showing how using metrics helps you make better business decisions.

Instrument the software itself so you can show actual usage and performance. Then the team can have conversations about which features are being used and which are not. Get into the habit of putting in this instrumentation from the start.

Use Dave McClure's Pirate Metrics (AARRR) to position conversations with the people for whom the team is building software. Ask questions like "What will this new feature help with?", "Does this change help to acquire new users or retain existing users?", and so on.

If you are working with more than one customer (whether internal or external), make them or their representatives aware of a shared value model and make them part of the process for deciding where and when to focus on which parts.

Metrics are about making change visible. That visibility of change might be incremental, towards or away from a goal, or it might be the first sign of a bigger change. Don't carry on doing something just because you can or because you have invested time and effort in it. Be part of the bigger business decisions that will determine future business success.

Terminology

AARRR Metrics: Acquisition, Activation, Retention, Referral, Revenue - a set of metrics defined by Dave McClure in 2007 to determine whether online services are performing well as a whole.

BDD: *see Behavior-driven Development*

Behavior-driven Development: an agile software development practice that encourages collaboration among developers, QA and non-technical or business participants in a software project. Popularized by Daniel Terhorst-North from 2006 onwards.

CFD: *see Cumulative Flow Diagram*

Cumulative Flow Diagram: metric to observe flow in our team and spot impediments, typically implemented by plotting the amount of work in each state of our process over a period of time.

Estimating: in the context of this book, it's the common style used by teams to predict how long a piece of work is going to take, relying on people's informed best-guess. Usually based on techniques like story points, velocity, t-shirt size, ideal days, or similar.

Flow: movement and delivery of customer value through a process (Vacanti2015). It's how consistently and continuously work moves through our process all the way to a customer.

Flow Efficiency: percentage of time that work spends being actively worked on, as opposed to the total amount of time the work takes, including wait time. A high flow efficiency means that work flows through our process as fast as it possibly can, with minimal wait time or delay.

Forecasting: the prediction of work completion, as well as many other business questions, based on the use of historical data to simulate what might happen in the future. The results are expressed as a list of possible outcomes with their likelihood.

Lead Time: the time that a piece of work takes to go from start (commitment point) to end (last state of influence) of a process. Note: we are using the names that seem to be most common in the Kanban community, but be aware that different people use different terms when meaning the same things. Most famously, Dan Vacanti calls it "cycle time" in his excellent book "Actionable agile metrics for predictability" (Vacanti2015).

Little's Law: states that $averageLeadTime = \frac{averageWIP}{averageThroughput}$. In simple terms it means that, on average, if you want stories to go faster (shorter lead time), you either have to increase the throughput (get more done) or reduce the amount of work in progress (work on less things at the same time).

Metrics: in the context of this book, the measure of some aspect of the team's work. They help us gather insights, answer questions and enable the team to make better business decisions. We split them into value, quality and operational metrics.

OEM: Original Equipment Manufacturer.

Quality: conformance to requirements, as defined by Philip B. Crosby's book "Quality is Free" (Crosby1987).

Pirate Metrics: *see AARRR Metrics* **PO**: Product Owner.

RAS model: Reliability, Availability, Serviceability - model developed by IBM (RAS) to look at different dimensions of quality.

Story Points: a measure of the perceived size of a user story. For example, something that is perceived as simple and quick to implement might be allocated a single story point, whereas something complex and long might be given ten story points.

Throughput: the number of work items that were completed during a particular period of time (e.g. in a sprint, or an iteration). Often also called delivery rate, or story count.

Value: why we do what we choose to do. Metrics for value help us focus on the things that matter, allowing us to not waste time and do something worthwhile.

Velocity: number of story points that the team completes in a sprint. Often used to make predictions about what the team can deliver, e.g. to decide how many stories they can include in the next sprint, or to determine how many sprints it will take to complete a given set of user stories.

WIP: *see Work In Progress*

Work In Progress: any piece of work that has been started but has not been completed yet. High amounts of work in progress typically results in a number of problems, including slower progress (Little's Law), more time wasted in context switching, and lower quality.

References and further reading

 Due to limitations of the LaTeX text processor used by Leanpub to generate the book from the manuscript files, some of the URLs in this section may not display correctly. A full list of references is available on the book website.

Visit BizMetricsBook.com for details.

Introduction

Adzic2014 - Gojko Adzic, 2014 'Zone of control vs Sphere of influence' 12-09-2014. [Online] https://gojko.net/2014/09/12/zone-of-control-vs-sphere-of-influence/ [Accessed 04-Feb-2018]

Anderson2010 - David J. Anderson, *Kanban: Successful Evolutionary Change for Your Technology Business*. Blue Hole Press, 2010.

Sinek2011 - Simon Sinek, *Start With Why: How Great Leaders Inspire Everyone To Take Action*. Penguin, 2011.

Chapter 1 - Throughput

AndersonCarmichael2016 - David J Anderson, Andy Carmichael, *Essential Kanban Condensed*. Blue Hole Press, 2016.

A3Thinking - Crisp, 'A3 Template'. [Online] https://www.crisp.se/gratis-material-och-guider/a3-template [Accessed 08-Feb-2018]

Brodzinski2015a - Pawel Brodzinski, 2015 'Story Points and Velocity: The Good Bits' 18-02-2015. [Online] http://brodzinski.com/2015/02/story-points-velocity-the-good-bits.html [Accessed 08-Feb-2018]

Burrows2014 - Mike Burrows, 2014 'The Kanban Survivability Agenda' 20-02-2014. [Online] https://www.infoq.com/articles/kanban-agenda-part3-survivability [Accessed 08-Feb-2018]

Carroll2016 - Ian Carroll, 2016 'No correlation between estimated size and actual time taken' 18-04-2016. [Online] https://www.iancarroll.com/2016/04/18/no-correlation-between-estimated-size-and-actual-time-taken/ [Accessed 08-Feb-2018]

Cohn2014 - Mike Cohn, 2014 'The Main Benefit of Story Points' 09-09-2014. [Online] https://www.mountaingoatsoftware.com/blog/the-main-benefit-of-story-points [Accessed 08-Feb-2018]

Cohn2016 - Mike Cohn, 2016 'What are Story Points?' 23-08-2016. [Online] https://www.mountaingoatsoftware.com/blog/what-are-story-points [Accessed 08-Feb-2018]

Cynefin - Dave Snowden, 'Cynefin Framework Introduction'. [Online] http://cognitive-edge.com/videos/cynefin-framework-introduction/ [Accessed 08-Feb-2018]

DeathMarch - 'Death march (project management)'. [Online] https://en.wikipedia.org/wiki/Death_march_(project_management) [Accessed 08-Feb-2018]

Duarte2012 - Vasco Duarte, 2012 'Story Points Considered Harmful - Or why the future of estimation is really in our past...'

25-01-2012. [Online] http://softwaredevelopmenttoday.blogspot.co.uk/2012/01/story-points-considered-harmful-or-why.html [Accessed 08-Feb-2018]

Duarte2015 - Vasco Duarte, *No Estimates.* Oikosofy Series, 2015. http://noestimatesbook.com/

Forss2016 - Hakan Forss, 2016 'KATA - Habits for lean learning Agile Australia 2016' 20-06-2016. [Online] https://www.slideshare.net/HkanForss/kata-habits-for-lean-learning-agile-australia-2016 [Accessed 08-Feb-2018]

GermanTankProblem - 'German tank problem'. [Online] https://en.wikipedia.org/wiki/German_tank_problem [Accessed 08-Feb-2018]

Maccherone2014 - Larry Maccherone, 2014 'The Impact of Lean and Agile Quantified: 2014' 26-01-2015. [Online] https://www.infoq.com/presentations/agile-quantify (around minute 22) [Accessed 08-Feb-2018]

MagennisSpreadsheets - http://bit.ly/SimResources contains many spreadsheet templates related to forecasting and modelling software projects, maintained by Troy Magennis.

PlanningPoker - 'Planning Poker'. [Online] https://www.agilealliance.org/glossary/poker/ [Accessed 08-Feb-2018]

Rao2014 - Mukesh Rao, 2014 'Estimation Using Ideal Days' 27-03-2014. [Online] https://www.scrumalliance.org/community/articles/2014/march/estimation-using-ideal-days [Accessed 08-Feb-2018]

Reinertsen2009 - Donald G. Reinertsen, *The Principles of Product Development Flow: Second Generation Lean Product Development.* Celeritas Pub, 2009. Chapter 5 in particular.

Rother2009 - Mike Rother, *Toyota Kata: Managing People for Improvement, Adaptiveness and Superior Results.* McGraw-Hill Education, 2009.

Savage2012 - Sam L. Savage, *The Flaw of Averages: Why We Under-*

estimate Risk in the Face of Uncertainty. John Wiley & Sons, 2012.

Singh2016 - Vikram Singh, 2016 'Agile Estimation Techniques' 18-01-2016. [Online] https://www.scrumalliance.org/community/articles/2016/january/agile-estimation-techniques [Accessed 08-Feb-2018]

Talai2014 - Nader Talai, 2014 'Does size matter?' 09-04-2014. [Online] http://cdn2.hubspot.net/hubfs/2495218/valueglide_dec2016_files/DOCs/AgileOrganisation_EstimationJourney.pdf [Accessed 08-Feb-2018]

ThoughtWorks2013 - Various authors, 2013 'How do you estimate on an Agile project?'. [Online] http://info.thoughtworks.com/rs/thoughtworks2/images/twebook-perspectives-estimation_1.pdf [Accessed 08-Feb-2018]

Vacanti2015 - Daniel S. Vacanti, *Actionable Agile Metrics for Predictability: An Introduction*. Daniel S. Vacanti, Inc, 2015.

Zheglov2014a - Alexei Zheglov, 2014 'Inside a Lead Time Distribution' 07-09-2014. [Online] https://connected-knowledge.com/2014/09/07/inside-lead-time-distribution/ [Accessed 08-Feb-2018]

Chapter 2 - Lead Time

AndersonCarmichael2016 - David J Anderson, Andy Carmichael *Essential Kanban Condensed*. Blue Hole Press, 2016.

Berardinelli - Carl Berardinelli, 'A Guide to Control Charts'. [Online] https://www.isixsigma.com/tools-templates/control-charts/a-guide-to-control-charts/ [Accessed 12-May-2018]

Brodzinski2012a - Pawel Brodzinski, 2012 'Slack Time' 10-05-2012. [Online] http://brodzinski.com/2012/05/slack-time.html [Accessed 08-Feb-2018]

Carroll2016 - Ian Carroll. 'Why You Might Be Wasting Your Time with Story Point Estimation'. Solutioneers (blog), 30 January 2020.

https://www.solutioneers.co.uk/why-you-might-be-wasting-your-time-with-story-point-estimation/

Ma2011 - Dan Ma, 2011 'When Bill Gates Walks into a Bar' 04-09-2011. [Online] https://introductorystats.wordpress.com/2011/09/04/when-bill-gates-walks-into-a-bar/ [Accessed 12-May-2018]

Novkov2017 - Alex , 2017 'Track Your Aging Work In Progress' 23-02-2017. [Online] https://kanbanize.com/blog/aging-work-in-progress/ [Accessed 12-May-2018]

Savage2012 - Sam L. Savage, *The Flaw of Averages: Why We Under-estimate Risk in the Face of Uncertainty*. John Wiley & Sons, 2012. For a shorter explanation by the same author see:
Sam L. Savage, 2002 'The Flaw of Averages' 11-2012. [Online] https://hbr.org/2002/11/the-flaw-of-averages [Accessed 18-May-2018]

Talai2014 - Nader Talai, 2014 - 'Does size matter?' https://cdn2.hubspot.net/hubfs/2495218/valueglide_dec2016_files/DOCs/AgileOrganisation_EstimationJourney.pdf

Vacanti2015 - Daniel S. Vacanti, *Actionable Agile Metrics for Pre-dictability: An Introduction*. Daniel S. Vacanti, Inc, 2015.

Vacanti2018 - Daniel S. Vacanti, 2018 'Actionable Agile Metrics with Daniel Vacanti' 14-01-2018. [Online podcast] https://podcast.agileuprising.com/actionable-agile-metrics-with-daniel-vacanti/ [Accessed 12-May-2018]

Chapter 3 - Forecasting and planning

Anderson2011 - David Anderson, 2011 'Predictability and Mea-surement with Kanban by David Anderson' 17-10-2011. [Online] https://vimeo.com/32436546 [Accessed 18-May-2018]

Bakardzhiev2014 - Dimitar Bakardzhiev, 2014 '#NoEstimates Project Planning Using Monte Carlo Simulation' 01-12-2014. [Online] https:

//www.infoq.com/articles/noestimates-monte-carlo [Accessed 21-Jun-2018]

Brodzinski2015b - Pawel Brodzinski, 2015 'Economic Value of Slack Time' 29-01-2015. [Online] http://brodzinski.com/2015/01/slack-time-value.html [Accessed 18-May-2018]

GermanTankProblem - 'German tank problem'. [Online] https://en.wikipedia.org/wiki/German_tank_problem [Accessed 08-Feb-2018]

Kingman Formula - Invistics, 2017, 'The King of Manufacturing Equations' 28-03-2017. [Online] http://www.invistics.com/the-king-of-manufacturing-equations/ [Accessed 18-May-2018]

Magennis2015 - Troy Magennis, 2015 'Agile 2015 - Risk - The Final Agile Enterprise Frontier' 06-08-2015. [Online] https://github.com/FocusedObjective/FocusedObjective.Resources/blob/master/Presentations/Agile%202015%20-%20Risk%20-%20The%20Final%20Agile%20Enterprise%20Frontier%20(Troy%20Magennis).pdf [Accessed 18-May-2018]

Public GitHub repository for Business Metrics book - https://github.com/SkeltonThatcher/bizmetrics-book. This has many useful spreadsheets to help you get started.

Reinertsen2009 - Donald G. Reinertsen, *The Principles of Product Development Flow: Second Generation Lean Product Development.* Celeritas Pub, 2009. Chapter 5 in particular.

Savage2012 - Sam L. Savage, *The Flaw of Averages: Why We Underestimate Risk in the Face of Uncertainty.* John Wiley & Sons, 2012. For a shorter explanation by the same author see: Sam L. Savage, 2002 'The Flaw of Averages' 11-2012. [Online] https://hbr.org/2002/11/the-flaw-of-averages [Accessed 18-May-2018]

Sunk Cost Fallacy - David McRaney, 2011 'The Sunk Cost Fallacy' 25-03-2011. [Online] https://youarenotsosmart.com/2011/03/25/the-sunk-cost-fallacy/ [Accessed 18-May-2018]

Vacanti2015 - Daniel S. Vacanti, *Actionable Agile Metrics for Predictability: An Introduction.* Daniel S. Vacanti, Inc, 2015.

Chapter 4 - Metrics for Flow

Anderson2013 - David Anderson, 2013 'Who is your Vice President of Delay?' 24-10-2013. http://djaa.com/who-your-vice-president-delay [Accessed 02-Dec-2018]

BlackSwanFarming - Black Swan Farming, 'Cost of Delay'. [Online] http://blackswanfarming.com/cost-of-delay/ [Accessed 02-Dec-2018]

Brodzinski2013 - Pawel Brodzinski, 2013 'Cumulative Flow Diagram' 15-06-2013. [Online] http://brodzinski.com/2013/07/cumulative-flow-diagram.html [Accessed 02-Dec-2018]

Burrows2014 - Mike Burrows, *Kanban from the Inside*. Blue Hole Press, 2014.

Deming1982 - W. Edwards Deming, *Out of the Crisis*. The MIT Press, 2000.

ForsgrenHumbleKim2018 - Nicole Forsgren PhD, Jez Humble and Gene Kim, *Accelerate: The Science of Lean Software and Devops: Building and Scaling High Performing Technology Organizations*. Trade Select, 2018.

Forss2013 - Hakan Forss, 2013 'LKUK13: The Red Brick Cancer' 03-12-2013. [Online] https://www.youtube.com/watch?v=sOb7Qqs2fe8 [Accessed 02-Dec-2018]

HumbleFarley2010 - Jez Humble, David Farley, *Continuous Delivery*. Addison Wesley, 2010.

Lowe2014a - David Lowe, 2014 'Little's Law' 28-10-2014. [Online] https://scrumandkanban.co.uk/littles-law/ [Accessed 02-Dec-2018]

Lowe2014b - David Lowe, 2014 'Little's Law and CFDs' 03-12-2014. [Online] https://scrumandkanban.co.uk/littles-law-and-cfds/ [Accessed 02-Dec-2018]

Lowe2015 - David Lowe, 2015 'Flow Efficiency' 01-06-2015. [Online] https://scrumandkanban.co.uk/flow-efficiency [Accessed 02-

Dec-2018]

MagennisSpreadsheets - http://bit.ly/SimResources contains many spreadsheet templates related to forecasting & modelling software flow, curated by Troy Magennis.

Reinertsen2009 - Donald G. Reinertsen, *The Principles of Product Development Flow*. Celeritas Pub, 2009. See Chapter 5 in particular.

Rothman2015 - Johanna Rothman, 2015 'Resource Efficiency vs. Flow Efficiency, Part 1: Seeing Your System' 13-09-2015. [Online] https://www.jrothman.com/mpd/agile/2015/09/resource-efficiency-vs-flow-efficiency-part-1-seeing-your-system/ [Accessed 02-Dec-2018]

Skarin2008 - Mattias Skarin, 2018 'Agile game – Pass the pennies' 08-09-2008. https://blog.crisp.se/2008/09/08/mattiasskarin/1220882915232 [Accessed 02-Dec-2018]

Vacanti2015 - Daniel S. Vacanti, *Actionable Agile Metrics for Predictability: An Introduction*. Daniel S. Vacanti, Inc, 2015.

Wester2016 - Julia Wester, 2016 'Flow Efficiency: A great metric you probably aren't using' 25-09-2016. http://brodzinski.com/2012/05/slack-time.html [Accessed 08-Feb-2018]

Yip2018 - Jason Yip, 2018 'Why T-shaped people?' 24-03-2018. [Online] https://medium.com/@jchyip/why-t-shaped-people-e8706198e437 [Accessed 02-Dec-2018]

Chapter 5 - Metrics for Quality

Abela2006 - Andrew Abela. 'Choosing a Good Chart'. Extreme Presentation (blog), 6 September 2006. https://extremepresentation.com/choosing_a_good-2/

Adzic2011 - Gojko Adzic, *Specification by Example: How Successful Teams Deliver the Right Software*. Manning Publications, 2011.

Big Buck Bunny - Wikipedia. 'Big Buck Bunny'. In Wikipedia, 27

August 2019. https://en.wikipedia.org/w/index.php?title=Big_Buck_Bunny&oldid=912659532

Cazaly2019 - Lynne Cazaly. Ish: The Problem with Our Pursuit for Perfection and the Life-Changing Practice of Good Enough. Lynne Cazaly, 2019.

CrispinGregory2009 - Lisa Crispin and Janet Gregory. *Agile Testing: A Practical Guide for Testers and Agile Teams*. Pearson Education, 2009. p.79

Crosby1987 - Philip B. Crosby. *Quality Is Free: The Art of Making Quality Certain*. New edition edition. New York u.a.: Penguin, 1987. p.15

Dettmer2007 - H. William Dettmer. *The Logical Thinking Process: A Systems Approach to Complex Problem Solving*. ASQ Quality Press, 2007.

Knaflic2015 - Cole Nussbaumer Knaflic. *Storytelling with Data: A Data Visualization Guide for Business Professionals*. John Wiley & Sons, 2015.

Magennis2017 - Troy Magennis. 'Data-Driven Coaching - Safely Turning Team Data into Coaching Insights'. 2017. https://www.infoq.com/presentations/data-coach/

Matts2013 - Chris Matts. 'Introducing Staff Liquidity (1 of n)'. The IT Risk Manager (blog), 24 November 2013. https://theitriskmanager.com/2013/11/24/introducing-staff-liquidity-1-of-n/

North2017 - Dan North. How to Fail with BDD | SkillsCast. Skillsmatter, 2017. https://skillsmatter.com/skillscasts/10695-how-to-fail-with-bdd

Reinertsen2009 - Donald G. Reinertsen, *The Principles of Product Development Flow: Second Generation Lean Product Development*. Celeritas Pub, 2009. p.55

RAS - Wikipedia. 'Reliability, Availability and Serviceability'. In Wikipedia, 27 January 2019. https://en.wikipedia.org/w/index.php?

title=Reliability,_availability_and_serviceability&oldid=880513040

Seddon2008 - John Seddon. Systems Thinking in the Public Sector. Triarchy Press, 2008.

Skelton2018 - Matthew Skelton. '5 Ways Site Reliability Engineering Transforms IT Ops'. TechBeacon, 9 October 2018. https://techbeacon.com/enterprise-it/5-ways-site-reliability-engineering-transforms-it-ops

Chapter 6 - Metrics for Value

Adzic2013 - Gojko Adzic. 'Scrum, Velocity, and Driving down the Motorway the Wrong Way'. Gojko's Blog (blog), 12 September 2013. https://gojko.net/2013/09/12/scrum-velocity-and-driving-down-the-motorway-the-wrong-way/

DeMarco1995 - Tom DeMarco. *Why Does Software Cost So Much?: And Other Puzzles of the Information Age.* New York, NY, USA: Dorset House Publishing Co., Inc., 1995. p43.

Drucker2012 - Peter Drucker. The Practice of Management. Routledge, 2012. p.29

Hubbard2014 - Douglas W. Hubbard. *How to Measure Anything Workbook.* John Wiley & Sons, 2014.

McClure2007 - Dave McClure. 'Startup Metrics for Pirates'. Business, 2007-08-08. https://www.slideshare.net/dmc500hats/startup-metrics-for-pirates-long-version

Perri2018 - Melissa Perri. *Escaping the Build Trap.* 1 edition. S.l.: O'Reilly, 2018.

Richer2017 - Julian Richer. *The Richer Way.* Cornerstone, 2017.

Young2014 - Chris Young. 'Down To The Wire', 2014. http://worldofchris.github.io/down-to-the-wire/

About the authors

Mattia Battiston

Mattia Battiston

Mattia is originally from Verona, the city of love and home of Romeo of Juliet. He is a software developer and team leader with a great passion for learning and continuous improvement.

His focus is to help teams strive to get better, using Kanban, Lean, Agile, and of course a lot of data and metrics.

He's been interested in everything to do with Agile since the beginning of his career in 2008. He loves attending and speaking at conferences and meetups for sharing experiences and learning from each other.

Chris Young

Chris Young

Chris has been a computerist since age 12 when he was introduced to the magic of the Commodore PET. He works with engineering and management teams with the aim of getting the best results possible for all involved.

The role and value of metrics in this came from discovering the Lean/Kanban community around 2010 which got him measuring things in earnest.

He is an active member of the Lean/Agile/DevOps community speaking at Meet Ups and conferences across Europe including GOTO Berlin, Agile Cambridge, CukeUp and QCon.

Why we wrote this book

Mattia: having experimented with data and metrics for a long time, I wanted to write this book to share what worked for me and help people learn from my mistakes. Metrics have massively helped me and my teams make better decisions, have better discussions, and improve every day. I hope that reading this book will help others achieve the same, if not more.

Chris: I wanted to share my success and failure in using metrics to improve the work and outcomes of the various teams I've been involved in. Having a foot in both the technology and business camps I have seen how shared metrics can provide a foundation for better communication and understanding and so lead to better outcomes. It's really important to me that this is a nurturing and helpful book and, like the work, it's very much a team effort.

Index

Conflux Books

Books for technologists by technologists

Our books help to accelerate and deepen your learning in the field of software systems. We focus on subjects that don't go out of date: fundamental software principles & practices, team interactions, and technology-independent skills.

Current and planned titles in the *Conflux Books* series include:

1. *Build Quality In* edited by Steve Smith and Matthew Skelton
2. *Better Whiteboard Sketches* by Matthew Skelton
3. *Internal Tech Conferences* by Victoria Morgan-Smith and Matthew Skelton
4. *Technical Writing for Blogs and Articles* by Matthew Skelton

Conflux Books also publishes the acclaimed *Team Guides for Software* collection: *Software Operability*, *Metrics for Business Decisions*, *Software Testability* and *Software Releasability*.

 Discover the *Conflux Books* series: confluxbooks.com

Lightning Source UK Ltd.
Milton Keynes UK
UKHW020621251121
394510UK00005B/26